Estuaries of the World

Series editor
Jean-Paul Ducrotoy

More information about this series at http://www.springer.com/series/11705

Tetsuo Yanagi
Editor

Eutrophication and Oligotrophication in Japanese Estuaries

The present status and future tasks

Springer

Editor
Tetsuo Yanagi
International EMECS Center
Kobe, Japan

ISSN 2214-1553 ISSN 2214-1561 (electronic)
Estuaries of the World
ISBN 978-94-017-9914-0 ISBN 978-94-017-9915-7 (eBook)
DOI 10.1007/978-94-017-9915-7

Library of Congress Control Number: 2015942917

Springer Dordrecht Heidelberg New York London

Printed on acid-free paper

Springer Science+Business Media B.V. Dordrecht is part of Springer Science+Business Media (www.springer.com)

Prologue

Human activities on land have expanded enormously since the industrialization became a main stream of economic development. World population which was estimated at 2.6 billion at the middle of the last century reached 6 billion at the end of the century and is expected to become 9.6 billion in 2050 according to medium-variant prospects of Population Division of DESA, United Nations. Furthermore, Gross World Product has increased from 7.0 trillion US dollar (2010, PPP) at 1950 to 77.2 trillion at 2010, more than ten times, according to Earth Policy Institute estimates. Most of the large-scale industries and big cities became located close to the sea, and the coastal sea environments have suffered ecosystem damages from the impacts of expanded human activities.

Japan achieved rapid economic expansion in the 1950s and 1960s, and gross domestic product (GDP) per capita increased almost a hundred times during 50 years starting from 1960. Behind the initial period of the economic growth, however, peoples met with various types of environmental deteriorations including large changes and pollutions in the coastal sea areas.

One example of severe industrial pollutions was found at the Dokai Bay in Kitakyushu City, northern part of Kyushu Island. Expansion of the steel work industries, followed by coal and heavy chemical industries' development around the bay caused severe marine water pollutions. No fish catch lasted for a long period, and at the end of the 1960s chemical oxygen demand (COD) at some part of the bay was reported as high as 70 mg/l resulting in the worst record of the polluted bays in the Asian region. The Dokai Bay has now completely renovated from being once called "The Sea of Death" through the hard endeavors and initiatives by the local community, labor unions, and local government. As a matter of fact, Kitakyushu was cited on the occasion of UNCED (Rio de Janeiro, 1992) as the capital of Sustainable Development of the World.

Economic developments are usually followed by urbanization, migration of population into big cities. Urban population concentrations are found in many countries worldwide, typically at big cities along coasts. Flat plains, if they exist behind inland seas or enclosed seas, provide convenient spaces for city development and attract people by supplying comfortable spaces and job opportunities.

The Tokyo Bay watershed area which covers Tokyo and six prefectures (15,500 km^2, 4% of the total area of Japan) has about 37 million inhabitants (30% of the total population), and economic activities in this area are expected to cover close to 40% of the total GDP of Japan. The dense population has big impacts to the Tokyo Bay water qualities through the number of rivers emptying into the bay. Natural beaches have been replaced by artificial concrete walls that separate land and sea. The water qualities once deteriorated have been improved mainly due to the implementation of the total pollutant load control system for nutrients to the enclosed seas. In these years, however, the blue tide events are still observed, which are considered to be caused by the deficit of dissolved oxygen in some parts of the bottom of the bay.

Also at the Seto Inland Sea, which is the largest enclosed sea in Japan, the drainage basin consists of 13 prefectures (18% of the total area of Japan). Population in the basin area is about 35 million (28% of the total population), and GDP in this area corresponds to 26% of the total as of 2008. Fishery has been the important industry in the middle part of the inland sea, but the water qualities were also impaired in the 1960s and 1970s when red tide occurrences were frequently observed. Again, the introduction of the total pollutant load control helped improvement of the coastal water qualities. Current nutrient concentrations have become so low that fishermen are worrying about lowering fish catches due to decreased amount of feeds, such as plankton in coastal waters.

Coastal seas, especially major estuaries in countries under industrial development stages, might face similar situations as what the estuaries in Japan went through. In order not to repeat the same sufferings, the lessons learned from the past events in Japan might be made the most of, which are covered comprehensively by this book. Analyses of coastal water environment affairs, management of human activities in backyards, governing systems including legislation and community activities, water-related technology development, and so on are quite dependent on local characteristics. Hopefully, all the estuaries are to be conserved as healthy spaces which provide full ecological services including sustainable supply of a variety of marine products, providing recreational, cultural, and spiritual values to the people living in the surroundings of estuaries and visitors.

President, International EMECS Center
Professor Emeritus, University of Tokyo
Motoyuki Suzuki

Contents

Contributors

Keita Furukawa Ocean Policy Research Institute, Sasakawa Peace Foundation, Minato-ku, Tokyo, Japan

Center for Oceanic Studies and Integrated Education, Yokohama National University, Hodogaya, Yokohama, Japan

Machiko Yamada Fukuoka Woman's University, Higashi-ku, Fukuoka, Japan

Tetsuo Yanagi International EMECS Center, Kobe, Japan

About the Editor

Dr. Tetsuo Yanagi is a coastal oceanographer. Tetsuo has about 500 scientific publications. He was awarded a Japan-France Oceanographic Society Award in 1986 and the Uda-Prize of Oceanographic Society of Japan in 2012. Tetsuo was a member of the Scientific Steering Committee of Land-Ocean Interactions in the Coastal Zone (LOICZ) and is a member of the Scientific Planning Committee of Japan's Environmental Management of Enclosed Coastal Seas (EMECS). He discovered the tide-induced residual current and proposed a new concept of Satoumi for integrated coastal sea management.

Professor of Emeritus
International EMECS Center, Kobe, Japan

Eutrophication and Oligotrophication in Japanese Estuaries: A Synthesis

1

Tetsuo Yanagi

Abstract

"Eutrophication and Oligotrophication in Japanese Estuaries", one in the series of books entitled "Estuaries of the World", addresses the question: What is the present states of Japanese estuaries from the viewpoint of eutrophication and oligotrophication? We describe three typical situations, namely, (1) eutrophication problems in Tokyo Bay, (2) oligotrophication problems in the Seto Inland Sea and (3) the disappearance of hypoxia in Dokai Bay. Total TP (total phosphorus) and TN (total nitrogen) loads control law has played an important role in these three bays. However, the results of the application of the law differ among the three bays as the characteristics of material cycling are different.

Keywords

Eutrophication • Hypoxia • Total loads control law • Oligotrophication • Habitat • ICM

1.1 Introduction

Eutrophication is a global environmental problem, and 726 sites located in coastal seas throughout the world are, or have recently been, suffering from related problems such as red tides, hypoxia and other biological problems. Among these sites, only 55 sites in the USA and the EU have shown a trend toward restoration (World Resources Institute 2012; http://www.wri.org/project/eutrophication/map). When submitted to eutrophication, ecological services provided by coastal seas are affected and preservation of coastal areas against eutrophication is recommended.

Japanese estuaries have suffered from serious environmental problems such as Toyama itai-itai disease (1955–2007) and Minamata disease (1956–present) from 1955 to 1970, a period of rapid economic growth in Japan. The average increase in gross national product (GNP) was 9.7 % during this period (Fig. 1.1). A wide variety of materials such as toxins, chemicals, nutrients and so on were emitted from the land without regulation during this time. Itai-itai disease is a bone disease caused by exposure to cadmium and first occurred among people along the Jintsu River, who lived on rice, vegetables and water polluted by cadmium from the Kamioka Mine Factory in Toyama Prefecture in 1955. Minamata disease is a nerve disease caused by methyl mercury emitted from a chemical factory and first occurred among fishermen who ate fish from Yatsushiro Bay in Kumamoto Prefecture in 1956.

1.2 Countermeasures Taken by the Japanese Government

The central government enacted a new water quality control law in 1970 to solve serious problems related to the environment, and the emission of toxic material from factories, including mercury and cadmium, has since been prohibited by applying a "zero-emission principle". Since that time, no serious environmental problems such as the Minamata disease have occurred in Japanese estuaries.

T. Yanagi (✉)
International EMECS Center, Kobe, Japan
e-mail: tyanagi@riam.kyushu-u.ac.jp

T. Yanagi (ed.), *Eutrophication and Oligotrophication in Japanese Estuaries*,
Estuaries of the World, DOI 10.1007/978-94-017-9915-7_1

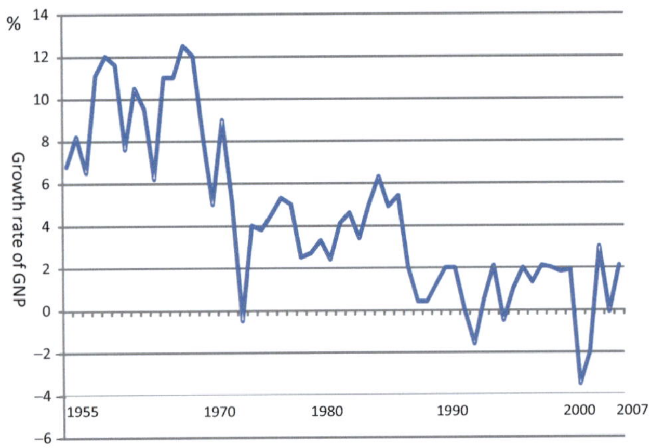

Fig. 1.1 Year-to-year variation in the Japanese gross national product (GNP)

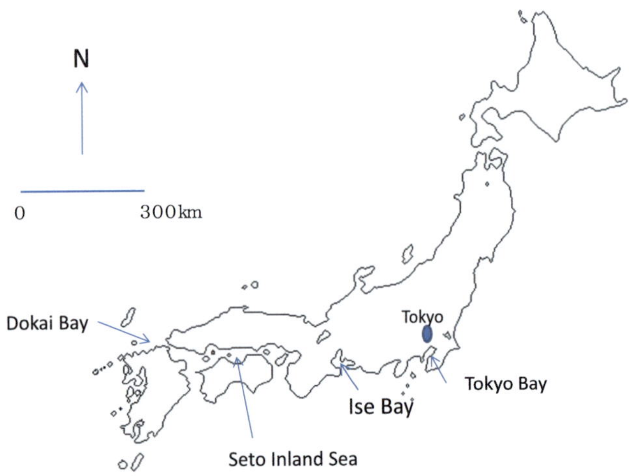

Fig. 1.2 A map of Japan showing Tokyo Bay, the Seto Inland Sea and Dokai Bay. Their geographical characteristics are shown in Table 1.1

However, eutrophication problems such as red tides and oxygen-deficient water masses, including hypoxia and anoxia, have become the main environmental problems in Japanese estuaries. During the period of rapid economic growth, nutrients concentrations in Japanese estuaries increased due to the rapid increase in phosphorus and nitrogen loads from the land. Central and local governments enacted the total nutrients loads control law in 1973 for semi-enclosed coastal seas such as Tokyo Bay, Ise Bay, the Seto Inland Sea and Dokai Bay (Fig. 1.2) because phosphorus and nitrogen, which are the main chemicals causing eutrophication, are not toxic but essential for biota including human beings. The same approach as adopted for toxic materials such as mercury and cadmium is not applicable to phosphorus and nitrogen.

1.3 Different Responses in Different Estuaries

Despite putting into force the total nutrients loads control law, nutrients concentrations in Tokyo Bay have not yet decreased due to excessive nutrients loads from land, and red tides and oxygen-deficient water masses continue to occur every summer even now. Nutrients concentrations in the Seto Inland Sea have not changed greatly, but the fish catch, including seaweed harvesting, is currently decreasing. On the other hand, nutrients concentrations in Dokai Bay have drastically decreased and hypoxia disappeared from the bay in 2011.

Regarding the definition of hypoxia and anoxia, there is, unfortunately, no general definition applicable to all estuaries. For example, hypoxia can be defined as when the dissolved oxygen (DO) concentration is lower than 2.5 ml/l (3.6 mg/l), when the life of benthos is threatened. Anoxia can be considered to occur when the concentration of DO is lower than 0.025 ml/l (0.036 mg/l) and when deoxidization by nitrates begins (Yanagi 1989). In this book, the definitions of hypoxia and anoxia are different for Tokyo Bay, the Seto Inland Sea and Dokai Bay because the environmental problems related to hypoxia differ in those estuaries, as will be discussed in Chaps. 2, 3 and 4.

The year-to-year changes in unit total phosphorus (TP) and total nitrogen (TN) loads (load/sea area) and yearly averaged TP and TN concentrations in the surface layer of Tokyo Bay, the Seto Inland Sea and Dokai Bay, which have a different spatial scale and population density in their catchment area (Table 1.1) and a different history of nutrients concentrations changes, are shown in Fig. 1.3.

Tokyo Bay has the largest population density in the watershed, the highest TN and TP loads per sea area and the highest TN and TP concentration now. Even now Tokyo Bay suffers from red tide and hypoxia each summer. The Seto Inland Sea has the lowest TN and TP concentrations among these three estuaries but the longest average residence time. Part of the Seto Inland Sea suffers from red tide and hypoxia every summer, but the Seto Inland Sea also suffers at the same time from a decrease in the fish catch due to oligotrophication. Dokai Bay had the highest TN and TP concentrations but these have been drastically decreased mainly due to the TN and TP loads control law. Furthermore, Dokai Bay has the shortest average residence time, the effect of loads control is easy to see, and hypoxia disappeared in 2011.

What has happened in these three selected Japanese estuaries and what the people did to prevent eutrophication in the coastal area of these estuaries during this period (1960–2010) are presented in this book.

Table 1.1 Geographical and hydro-dynamical characteristics of Tokyo Bay, the Seto Inland Sea and Dokai Bay

	Length (km)	Width (km)	Average depth (m)	Area (km^2)	Watershed area (km^2)	Population density in the watershed (km^{-2})	Tidal range in spring tide (cm)	Average residence time of fresh water (days)
Tokyo Bay	50	30	15	922	7,600	3,300	100	30
Seto Inland Sea	450	20	30	2,300	16,000	1,200	300	400
Dokai Bay	10	1	10	10	89	2,000	50	10

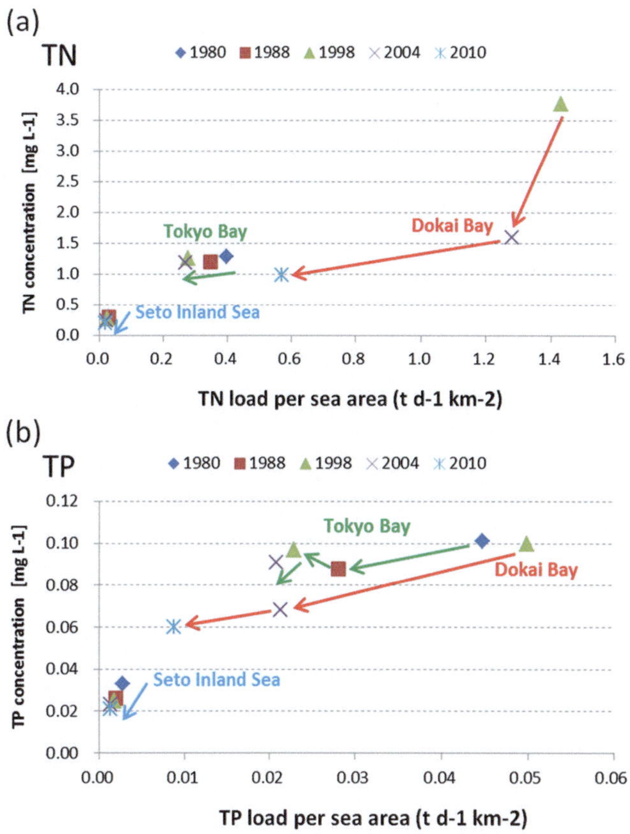

Fig. 1.3 Unit TP and TN loads and TP and TN concentrations in the surface layer of Tokyo Bay, the Seto Inland Sea and Dokai Bay

First of all, problems related to eutrophication in Tokyo Bay are described. Next, problems related to eutrophication and oligotrophication in the Seto Inland Sea are discussed. Finally, the process of hypoxia disappearing from Dokai Bay will be discussed.

We have learned many things from the experiences in Tokyo Bay, the Seto Inland Sea and Dokai Bay. In such coastal seas of Japan, red tides and hypoxia occurred as a result of eutrophication. Hypoxia due to eutrophication is a most serious phenomenon because it directly kills the benthos, damages the reproduction of zooplankton and results in the loss of comprehensive material cycling and a decrease in fish resources (Diaz and Rosenberg 2008).

In order to prevent the occurrence of red tides and hypoxia due to eutrophication, we first have to reduce TP and TN loads from the land. The total TP and TN loads control law was highly effective in Tokyo Bay and Dokai Bay because the ratio of land-derived TP and TN in the existing TP and TN in bays was very high there. It was particularly effective in Dokai Bay, where TP and TN loads decreased by 82 % and 60 %, respectively, from 1998 to 2011.

However, such a law was not so effective in the case of the Seto Inland Sea because the ratio of land-derived TP and TN was not high there (Fig. 3.38, Chap. 3).

1.4 Healthy and Comprehensive Material Cycling

Reducing TP and TN loads from land is not sufficient to prevent hypoxia because healthy and comprehensive material cycling is impossible without the existence of appropriate habitats for marine biota. Ecosystem-based management (EBM) is very important because biochemical processes are more important than physical processes for healthy and comprehensive material cycling in estuaries (Yanagi 2007). Therefore, we have to restore habitat conditions for marine biota in order to establish the comprehensive material cycling of phosphorus and nitrogen and to prevent the occurrence of red tides and hypoxia (Mee 2006). Habitat conditions differ from place to place in the coastal seas because they have high heterogeneity. As a result, we have to look for the best way to restore the habitat environment for marine biota by adaptive management, that is, the step-by-step process of countermeasures is very important and effective (adaptive management). For example, an artificial tidal flat was created in Tokyo Bay, eel grass beds were restored in the Seto Inland Sea, and the best method for the culture of bivalves was investigated using an experimental raft in the central part of Dokai Bay.

In order to stop oligotrophication in the Seto Inland Sea, we first have to restore the comprehensive material cycling of TP and TN and reconstruct a healthy ecosystem by restoring tidal flats, seagrass beds, sea-algae beds and shallow areas with a depth of light that allows sufficient penetration for marine biota.

1.5 Future Tasks

We need to continue environmental monitoring for detecting change and adaptive management. We have to clarify the necessary human interaction for establishing a rich coastal sea area not only for marine biota but also for the well-being of humans. Therefore, a wide variety of activities related to habitats creation and water purification have been conducted in Tokyo Bay, the Seto Inland Sea and Dokai Bay based on the participation of different stakeholders, including fishermen, local citizens, managers, scientists and so on. We have to promote such activities and organize them as ICM (Integrated Coastal Zone Management). The Satoumi concept is a very promising example of successful ICM (Yanagi 2010).

References

Diaz RJ, Rosenberg R (2008) Spreading dead zones and consequences for marine ecosystem. Science 321:926–929. doi:10.1126/science.1156401

Mee L (2006) Reviving dead zone. Scientific American, November 2006, 79–86

Yanagi T (1989) A summary of "symposium on oxygen-deficient water mass". Jpn Coast Oceanogr 26:141–145 (in Japanese with English abstract and captions)

Yanagi T (2007) Satoumi-a new concept for coastal sea management. TERRAPUB, Tokyo, 105p

Yanagi T (2010) Japanese commons in the coastal seas – how the satoumi concept harmonizes human activity in coastal seas with high productivity and diversity. Springer, New York, 113p

Eutrophication in Tokyo Bay

2

Keita Furukawa

Abstract

Tokyo Bay, located in the centre of Japan, has a temperate, humid climate. The bay is an enclosed environment, which is heavily populated and densely used. It is approximately 50-km long and 20-km wide and has an average depth of 15 m. Since 1950, increase in the concentration of population and industry in this river basin has caused radical changes in the coastal areas of Tokyo Bay. Some of these changes were eutrophication and a decline in the area of shallow sandy flats in the bay. As a result of eutrophication, Tokyo Bay has been affected by red tides (algal blooms) approximately 50 times a year and by blue tides (upwelling of hypoxic bottom water) approximately 3 times a year. The area of tidal flats has decreased by approximately 90 % over the last 100 years. These changes can affect the distribution of living organisms in the bay. The iconic richness of biological production in the shallow and tidal flats is under the threat of disappearance. On 26 March 2003, the Tokyo Bay Renaissance Promotion Council endorsed the 'Action Plan for Tokyo Bay Renaissance'. The goal is to restore the beautiful coastal environment for people to enjoy and sustain its natural biodiversity. The challenge of restoration is just beginning. Important and practical future aims are as follows: (1) to sustain the monitoring campaign with the cooperation of various stakeholders; (2) to create and maintain habitat for benthos, sessile organisms and fish larvae using an ecosystem approach; and (3) to establish public participation mechanisms.

Keywords

Eutrophication • Enclosed bay • Circulation • Red tide • *Edo-mae* culture • Renaissance project

2.1 Introduction

2.1.1 Tokyo Bay as an Estuary

Tokyo Bay is located in the centre of Japan between latitude 35 00′N and 35 40′N and longitude 139 40′E and 140 05′E (Fig. 2.1). It has a temperate, humid climate. The lowest temperature is approximately 5 °C during January and February, with the highest temperature of approximately 30 °C during July and August. The least amount of precipitation is observed in January and February. Heavy precipitation is observed during the rainy season in June, as well as during September and October when typhoons often hit Japan. Monthly precipitation is approximately 50 mm and 150 mm in January and September, respectively (Furukawa and Okada 2006).

The freshwater run-off from the catchment area of Tokyo Bay discharges into the bay on the order of 8–12 × 10^{12} tonnes/year. There are two major watersheds, namely, the Edo and the Ara River watersheds (Fig. 2.2). These rivers

K. Furukawa (✉)
Ocean Policy Research Institute, Sasakawa Peace Foundation, 3-4-10 Toranomon, Minato-ku, Tokyo 105-0001, Japan

Center for Oceanic Studies and Integrated Education, Yokohama National University, 79-5 Tokiwadai, Hodogaya, Yokohama 240-8501, Japan
e-mail: k-furukawa@spf.or.jp; keitaf@ynu.ac.jp

T. Yanagi (ed.), *Eutrophication and Oligotrophication in Japanese Estuaries*, Estuaries of the World, DOI 10.1007/978-94-017-9915-7_2

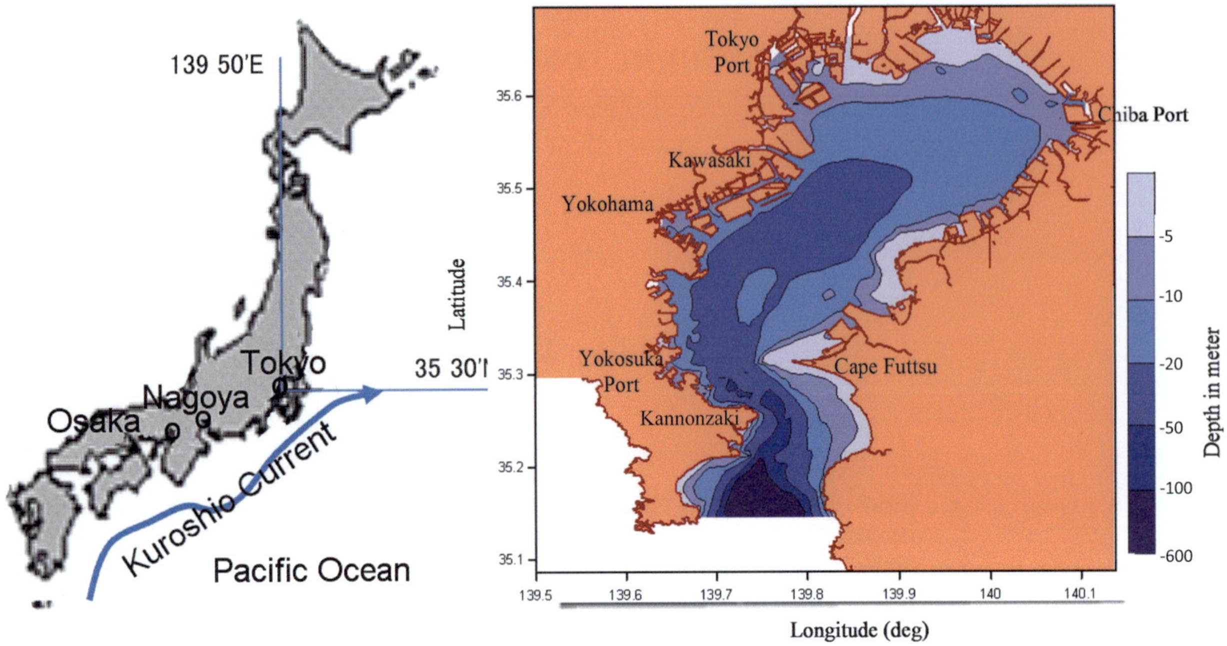

Fig. 2.1 Tokyo Bay

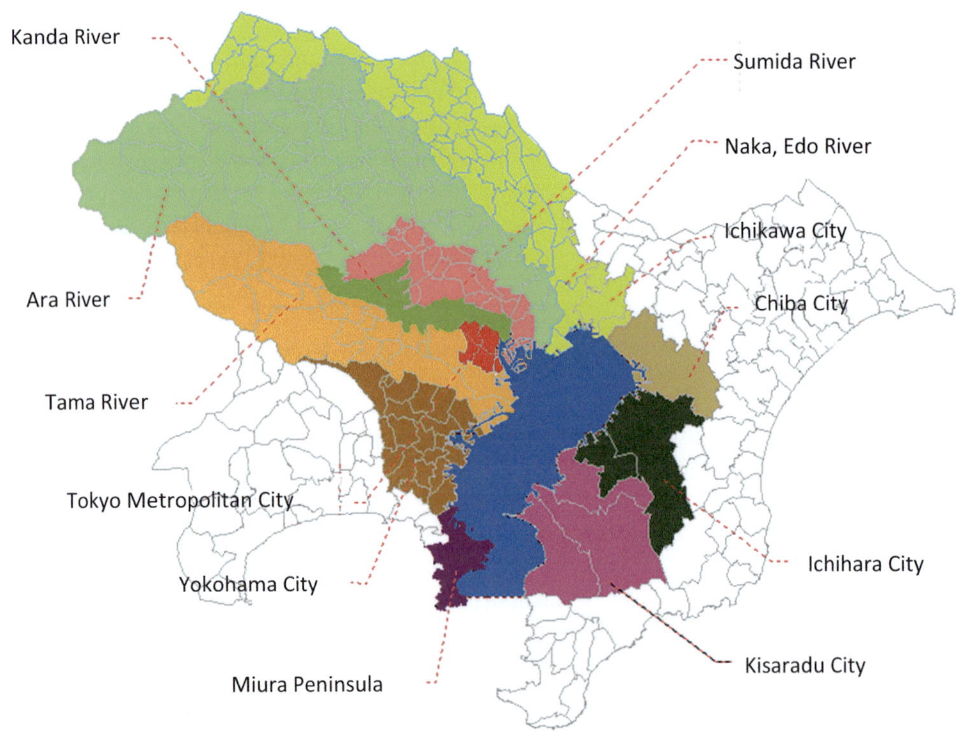

Fig. 2.2 Watersheds of Tokyo Bay

provide 50 % of the total freshwater supply of the bay. This water discharge forms a clear estuarine circulation from the head of the bay towards its mouth.

The currents in the bay are mainly caused by tides, density gradients, wind stress and input of oceanic water.

The tides are typically semidiurnal, with tidal ranges of approximately 1.5 m during the spring tide and approximately 0.5 m during the neap tide. The maximum tidal current is approximately 1.2 m s^{-1} between Cape Kannonzaki and Cape Futtsu and 0.2 m s^{-1} in the centre of the bay (Fig. 2.1). In general, the residual current forms a

Table 2.1 Catchments: Tokyo Bay, San Francisco Bay and Chesapeake Bay

	Units	Tokyo Bay	San Francisco Bay	Chesapeake Bay
Surface area	km^2	960[a]	1222[b]	18130[b]
Catchment area	10^3 km^2	7.6[a]	156.0[b]	166.0[b]
Average depth	m	15[a]	5[b]	6[b]
Population in catchment area (in 2000)	10^6 persons	27.8	10.0[b]	15.7[b]
Population per catchment area	10^3 persons km^{-2}	3.7	0.06	0.1
Population per surface area	10^3 persons km^{-2}	29	8.1	0.9
Reclamation area	km^2	249 (24.9 %)	240 (19.4 %)	12 (0.1 %)
Annual shipping freight containers (in 2000)[c]	10^6 TEU	5.8	1.9	1.6

[a]Ogura, 1993
[b]International EMECS Center, 2003
[c]Degerlund, 2005

clockwise circulation in the inner bay during summer, but this highly depends on wind direction and speed (Suzuki et al. 1997).

The area around Tokyo Bay is enclosed and is the most heavily populated and densely used bay area in Japan. The outer bay is open to the Pacific Ocean. The inner bay measures approximately 50 km from northeast to southwest and is 20-km wide with an average depth of 15 m (Fig. 2.1). There is no sill-type structure at the mouth of the bay; the bay is directly connected to the deep sea floor through a narrow continental shelf by a deep sea valley. The deepest part of the valley at the mouth of the bay has a water depth of 600 m, which allows the intrusion of oceanic currents directly into the bay (Hinata 2006). A total of 26 million people reside in the 7500 km^2 of the inner bay catchment area with a population density of 3,700 persons km^{-2}. Since 1950, the concentration of population and industry in this river basin has caused radical changes to the coastal areas of Tokyo Bay. Some of these changes were eutrophication and the loss of shallow sandy flats in the bay.

The concentrations of population and industry in the catchment area of Tokyo Bay are very high according to estimates based on the population per surface area (= population in the catchment area/surface area of the bay) and the total annual amount of freight at all ports in the bay. The population per surface area of Tokyo Bay is 29.0×10^3 persons km^{-2}, while the population per surface area of San Francisco Bay and Chesapeake Bay in the United States of America is 8.1×10^3 persons km^{-2} and 0.9×10^3 persons km^{-2}, respectively (Table 2.1). In addition, the amount of annual shipping freight containers in Tokyo Bay is 5.8×10^6 TEU[1] per year, whereas the amount in San Francisco Bay and Chesapeake Bay are 1.9×10^6 and 1.6×10^6 TEU per year, respectively.

There are several reasons for the concentration of population and industry in the catchment area of Tokyo Bay: (1) There are many suitable locations for ports in Tokyo Bay, where wave conditions are calm. (2) The catchment area of Tokyo Bay includes a wide plain, i.e. the Kanto Plain. (3) The hinterlands of the bay are suitable for industry, which involves the import of resources and export of products via sea. (4) The capital of Japan is located in the catchment area. (5) The climate of the catchment area is conducive to habitation.

[1] Twenty-foot equivalent unit: unit of trading volume normalised by the volume of a single twenty-foot container.

2.1.2 Topographical Features

Tokyo Bay was a continental shelf 120,000 years ago and became a plane incised by the deep valley of the Old Tokyo River approximately 20,000 years ago during the last glacial period. Relative sea level rise 6,000 years ago led to the initial formation of Tokyo Bay by inundation. Finally, sedimentary processes resulted in the present form of Tokyo Bay (after Kaizuka 1993: Fig. 2.3). Thus, Tokyo Bay is a sedimentary environment. In the 1950s, the bay was enclosed by an oval-shaped, smooth coastline with fringing tidal flats of about 2–6-km wide, and the nearshore bottom sediment was mainly sandy material. This has changed in recent decades, which includes significant alterations in the coastline and sediment fraction of the bay seafloor. At present, the coastline is discretised by landfill in the shallows and channel dredging. Consequently, major parts of the tidal flats and shallows have been altered. Consequently, the surface area of Tokyo Bay decreased by 26 % from 1900 to 2000. The present tidal flat area of Tokyo Bay is only 10 km^2 compared to 136 km^2 in 1900 (Fig. 2.6).

2.1.3 *Edo-mae* Culture

As shown in Fig. 2.4, Tokyo Bay has been well developed in terms of fisheries. The figure shows a catchment map drawn in 1908 overlain on a geological map from the 1800s. Many shallow banks and seagrass meadows were captured in the map. This map was part of a series of fisheries research reports in 1900. Subsequently, fisheries rights were devised in 1909 by the Meiji Fisheries Act; there is no indication of zoning of fisheries rights on the map. This indicates how people wisely used Tokyo Bay.

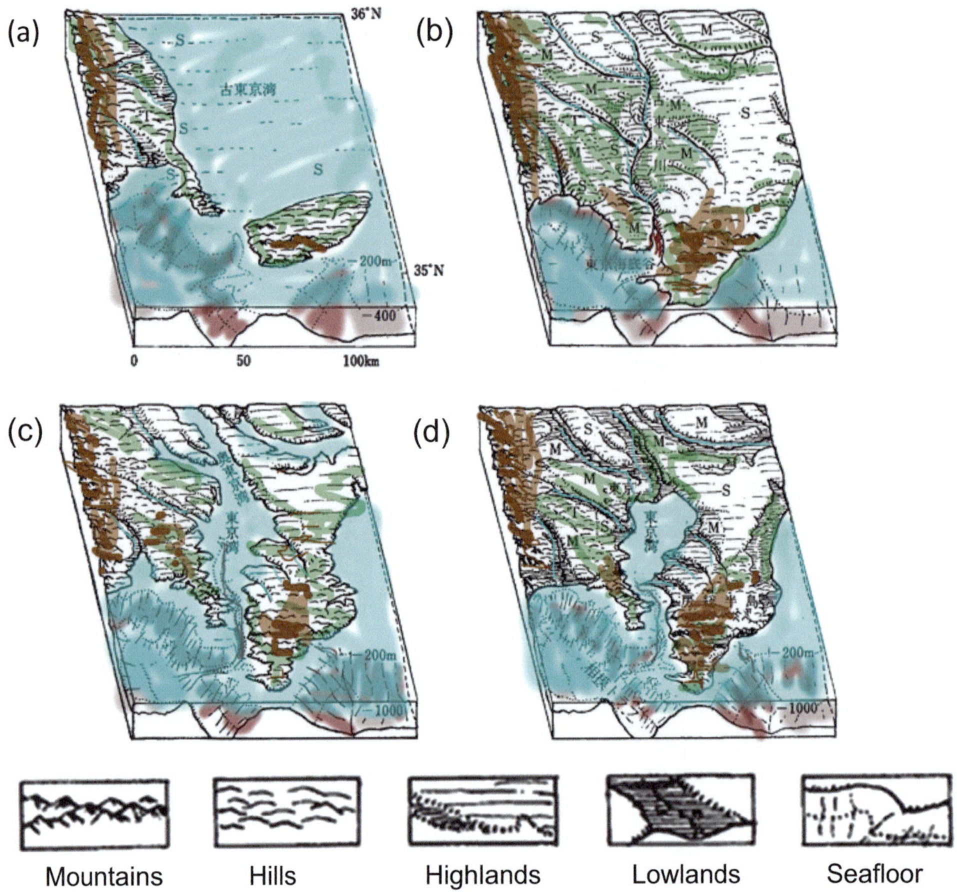

Fig. 2.3 Geological changes in Tokyo Bay. (**a**) 120,000 years ago, final interglacial period; (**b**) 20,000 years ago, final glacial period; (**c**) 6,000 years ago, after the glacial period; (**d**) twentieth century, present state (After Kaizuka 1993)

Edo-mae is a term used for seafood caught in front of *Edo* Metropolitan City. '*Mae*' implies both 'in front' and 'advance'. When an inbound ship arrived in *Edo* from the southwest shore of Tokyo Bay where the bay mouth opens, the cities located in that area were considered to be 'advanced' in *Edo* City. Thus, some people claim that *Edo-mae* could be extended down to the Tsurumi River. At present, *Edo-mae* is extended to the entire Tokyo Bay area (Fishery Agency 2005).

Typical *Edo-mae* are *nori* (Nori seaweed), *asari* (short-necked clam), *hamaguri* (clam), *unagi* (eel), *haze* (goby), *ebi* (shrimp), *shako* (squilla), *hirame* (flat fish), *suzuki* (sea bass), *saba* (mackerel) and *aji* (horse mackerel). These are said to be *Edo-mae sushi* materials. These catch were part of the rich seafood culture in the *Edo* era.

2.2 Environmental Changes in Recent Centuries

2.2.1 Changes in Population and Land Use

The population of prefectures facing Tokyo Bay was 10 million in the 1920s, which has increased four times (40 million); the population of the watershed was 26 million in 2000 (Fig. 2.5). The increasing population caused changes in land use, e.g. agricultural areas were altered to residential areas. Interestingly, paddy fields have not been used as dry fields because the alteration of paddy fields is difficult due to regulations. The Establishment of Agricultural Promotion Regions Act makes the alteration of paddy fields difficult. Paddy fields are mostly classified as urbanisation-controlled areas (Yoshikawa and Motonaga 2006).

The centres of the urbanised areas are located on river deltas or reclaimed lands. Thus, most major rivers (e.g. the Ara River and the Edo River) have been diverted by artificial channels.

2.2.2 Landfill and Shoreline Change

Figure 2.6 illustrates the changes in the bay from the 1950s to 2000. In the 1950s, the bay had an oval, smooth coastline with a fringing tidal flat (2–6-km wide), and the seafloor of the shoreward area was covered by sand. At present, the coastline has become eroded, and the seafloor of this region is covered by sludge with a high moisture content (weight of water/weight of sediment $\times$ 100) of over 200.

The direct causes of the change in sedimentary conditions are considered to be the reclamation of sandy areas and

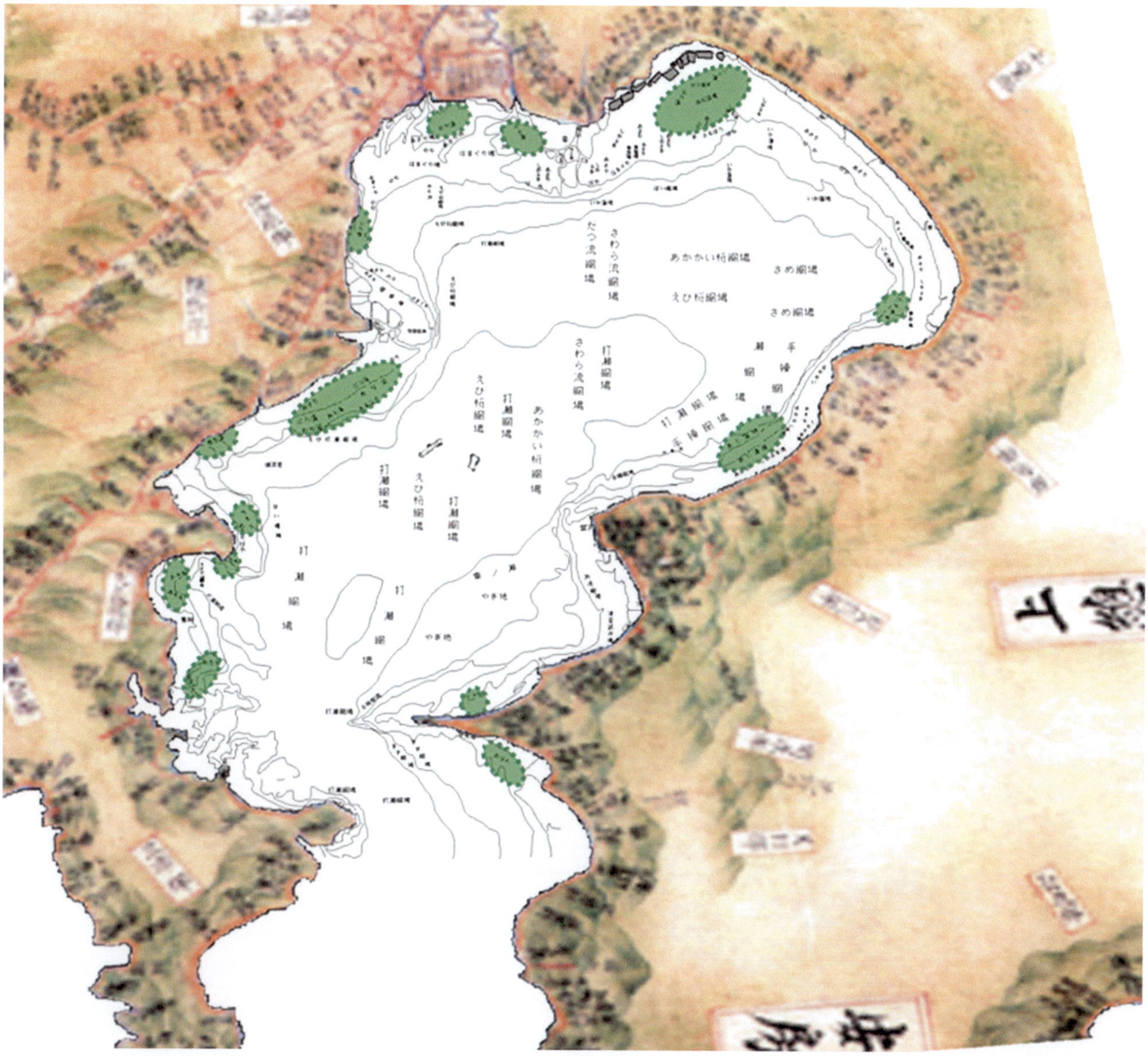

Fig. 2.4 Fisheries map of 1908 superimposed on a geographical map from the 1800s

heavy accumulation of organic matter. These changes are in turn driven by changes in bay circulation due to reclamation because the surface area was reduced by 80 % from 1960 to 2000. This reduction in surface area resulted in an 11 % decrease in the tidal range of the M_2 tidal component (Unoki and Konishi, 1999). In addition, the tidal current at the mouth of the inner bay reduced by 20 % from 1968 to 1983 (Yanagi and Onishi 1999).

2.2.3 Fisheries

Fisheries in Tokyo Bay possibly originated in the Jomon era (12000–2500 BC). Nevertheless, the major development started during the *Edo* era (1603–1868 CE). The necessity of fish supply for the Shogun family of Tokugawa and the emerging citizens of *Edo* promoted fisheries. Shogun Ieyasu Tokugawa recruited fishermen from western Japan to *Edo* and simultaneously provided support to existing local fishermen. *Nori* mariculture started at this point and become a major production of Tokyo Bay (called *Edo-mae*).

During the 260 years of the *Edo* era, there were many conflicts between fishermen. As a result, in 1816, an Edo bay fisheries protocol was established with the cooperation of five major fisheries parties in the bay, which was endorsed by 44 local fishing communities. According to the protocol, the shallow nearshore area was exclusively used by the local community, and the deeper offshore area was shared as a common resource (Committee of editing board for historical record of Tokyo Inner Bay 1971).

The total annual fish catch was more than 130,000 tonnes in the 1950s, approximately 90 % of the catch being shellfish. The catch started to gradually decline after 1960, reaching less than 50,000 tonnes per year in the 1970s. The main factor for this decline was the decrease in the shellfish catch. Since 1970s, the catch has continued to decline. In the 2000s, the total annual fish catch was 20,000 tonnes per year, which is less than one-sixth of the maximum catch in Tokyo Bay (Tokyo Bay Marine Environment Research Committee 2011; Fig. 2.7).

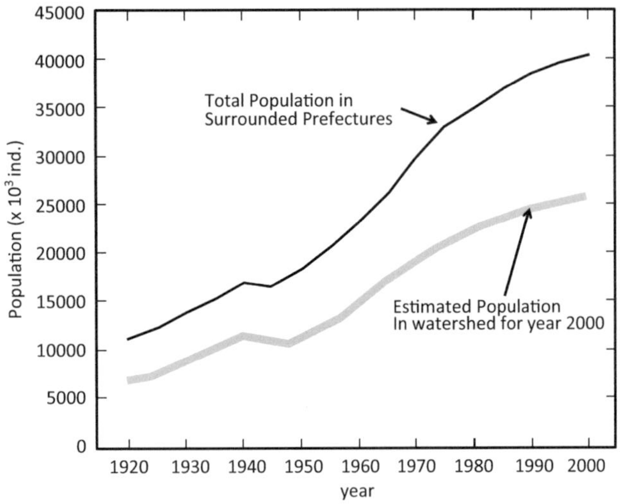

Fig. 2.5 80 years of population change in the Tokyo Bay watershed (Yoshikawa and Motonaga 2006)

The major factors causing the decline in the shellfish catch are considered to be the (1) decrease of habitat in the tidal flat areas and shallow-water regions because of reclamation; (2) degradation of the shellfish habitat due to anoxic water, red tides and blue tides; (3) competition with newly introduced species; (4) disease; and (5) destruction of the tidal flat network for benthic larvae transport (Tokyo Bay Marine Environment Research Committee 2011).

2.3 Status of the Bay

2.3.1 Circulation of the Bay

The current residence time of seawater in the inner Tokyo Bay is lower than that in the past. The residence time of seawater for the inner bay in 2002 was 20 days during

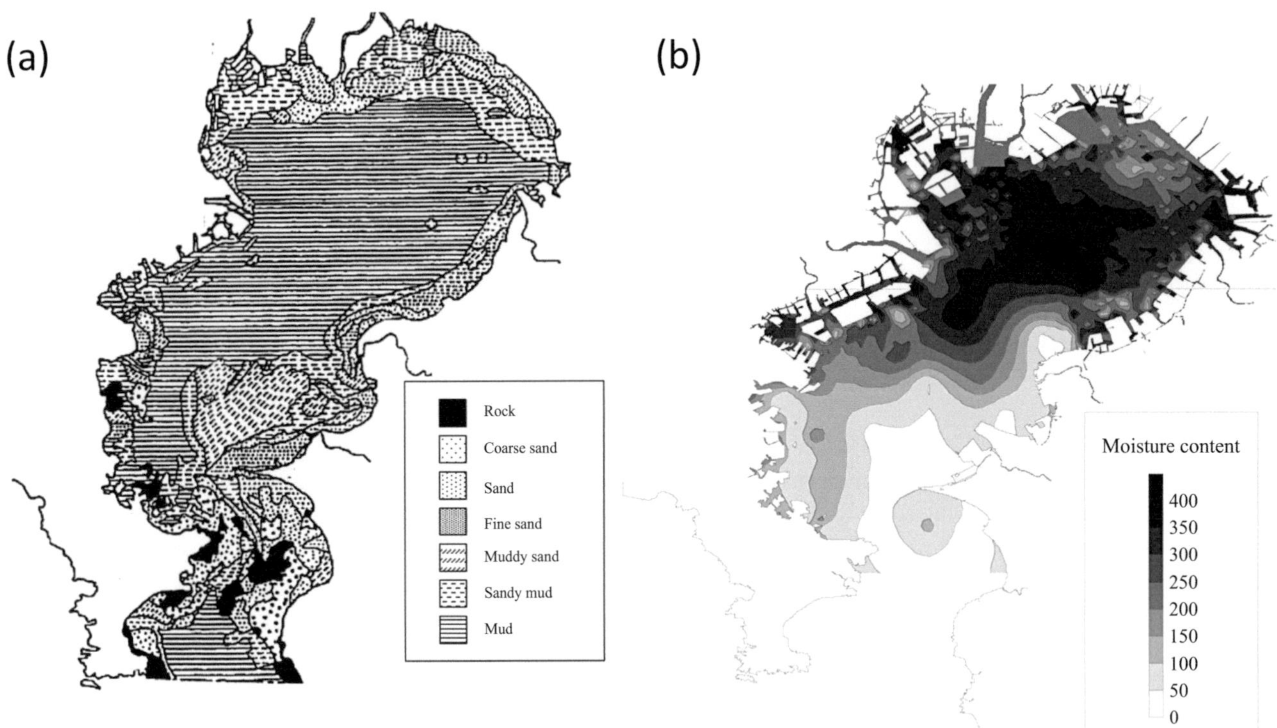

Fig. 2.6 Distribution of sediment in Tokyo Bay. (**a**) Main sediment fraction in the 1950s and (**b**) sediment moisture (%) in the 2000s (higher moisture is correlated with a richer mud fraction) (Furukawa and Okada 2006; Kaizuka 1993)

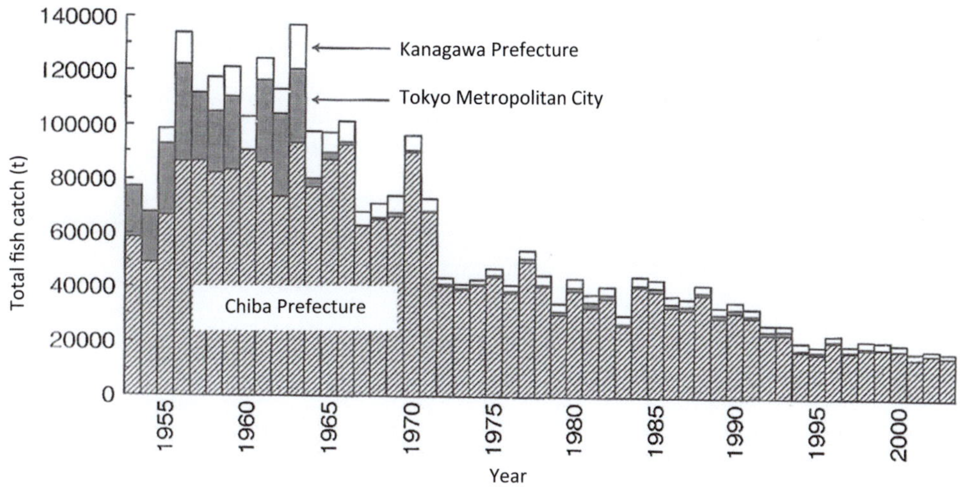

Fig. 2.7 Total fish catch in Tokyo Bay (Tokyo Bay Marine Environment Research Committee 2011)

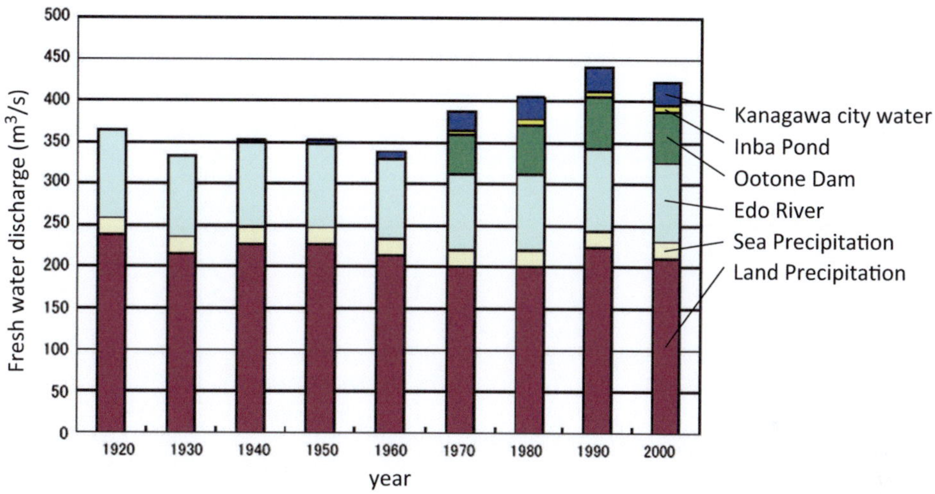

Fig. 2.8 Freshwater discharge into Tokyo Bay (1920–2000) (Takao et al. 2004)

summer and 40 days during winter (Takao et al. 2004). In contrast, the residence time of seawater in the 1960s was 1 month during summer and 3 months during winter (Unoki and Kishino 1977). Numerical simulations have indicated that the reduction in the residence time of seawater was mainly due to an increase in the freshwater input because of human population importing water from other catchment areas as well as the decrease in the surface area of the bay. The mechanisms of the reduction in the residence time of seawater are as follows: (1) Estuary circulation caused by freshwater input is the main cause of seawater exchange in Tokyo Bay (Unoki 1998). Consequently, the increase in freshwater input promotes estuary circulation and enhances seawater exchange. (2) The decrease in water surface area has caused a reduction in the tidal current speeds. The reduction in tidal currents weakens vertical mixing and strengthens stratification in Tokyo Bay. Thus, estuary circulation is promoted and seawater exchange is enhanced (Hinata and Furukawa 2006).

The freshwater input into Tokyo Bay increased by approximately 50 m^3s^{-1} from 1960 to 1970, which was approximately 350 m^3s^{-1} before 1960 and 400 m^3s^{-1} after 1970 (Fig. 2.8; Takao et al. 2004). The increase in freshwater input was due to an increase in the imported water mass from neighbouring catchments (Kanagawa City water, Inba Pond and Ootone Dam, as shown in Fig. 2.8) during the period 1960–1970. The imported water mass was used as city water and industrial water in the catchment area of Tokyo Bay.

The effects and patterns of oceanic water (typically the Kuroshio warm-water current) intrusion into Tokyo Bay are described by Hinata (2006). As shown in Fig. 2.9, during periods of oceanic water intrusion into the middle layer, the Kuroshio front approached the Uraga Channel. Nevertheless, judging by the salinity, the water mass was considered to have originated from river water. This mass was possibly formed as a result of the inflow of a large amount of freshwater and the approach of the Kuroshio front, which was affected by subsequent surface mixing. It remains difficult to

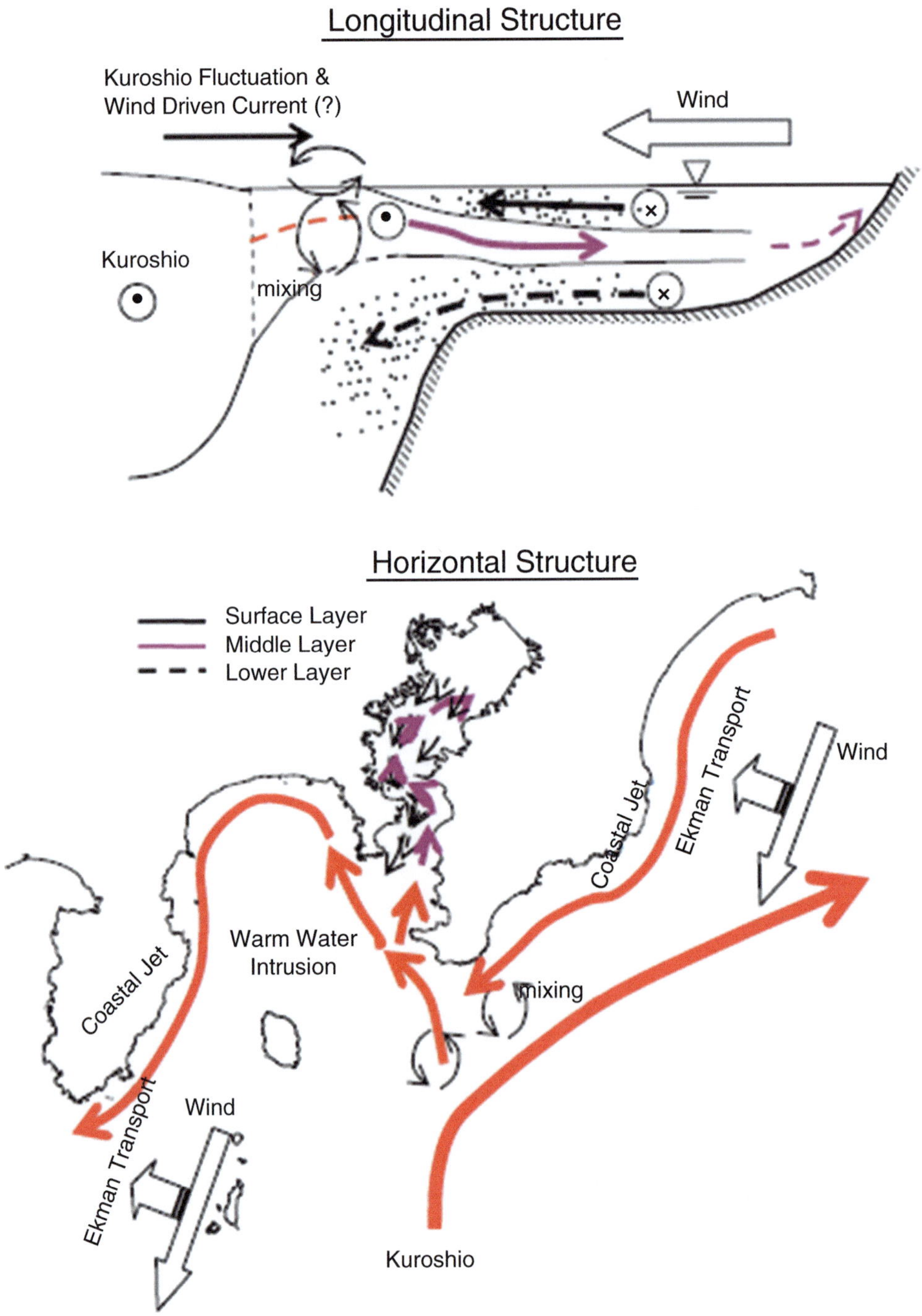

Fig. 2.9 Schematic diagram of flow structure during oceanic warm-water intrusion into the middle layer of Tokyo Bay (Hinata 2006)

obtain a clear picture of complex mass transport due to oceanic water intrusion.

Yanagi (2006) described how a significant amount of nitrogen (N) and phosphorus (P) standing stock in the bay originates from land. Because substantial amounts of both N and P are recycled within the bay, a reduction in inflow load does not cause direct and instant changes in the N and P concentrations in the bay. Yanagi (2006) quantitatively supported his results by showing the contribution of nutrients originating from the ocean. As shown in Sect. 4.4.2 (Fig. 2.37), the amounts of total nitrogen (TN) and total phosphorus (TP) are 19 % and 22 %, respectively.

2.3.2 Trophic Level of the Bay

Because of eutrophication, Tokyo Bay has been affected by red tides (algal blooms) a maximum of about 70 times a year and by blue tides (upwelling of hypoxic bottom water) a maximum of about 10 times a year (Fig. 2.10). Furthermore,

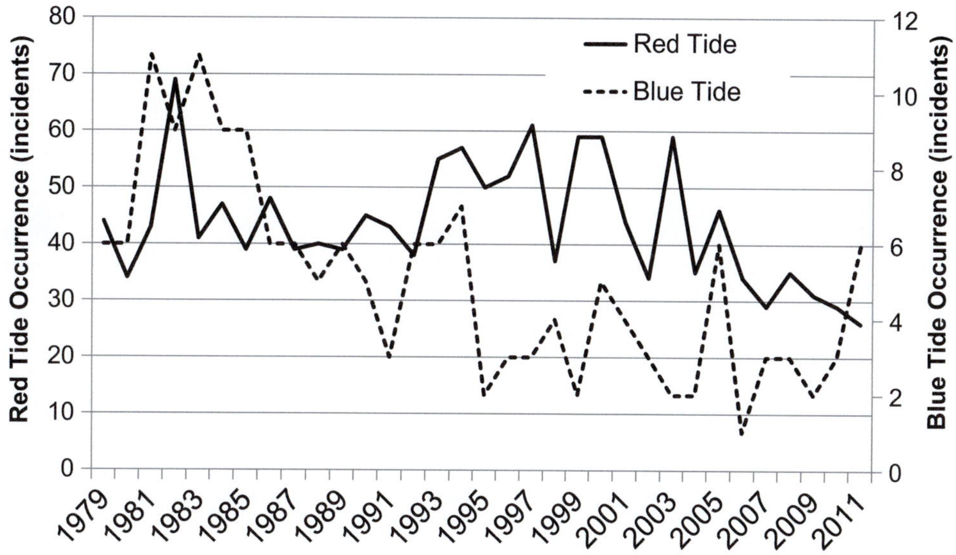

Fig. 2.10 Numbers of red tide and blue tide occurrences in Tokyo Bay between 1979 and 2011 (From Ministry of Environment 2010; Tokyo Bay Renaissance Promotion Council 2013)

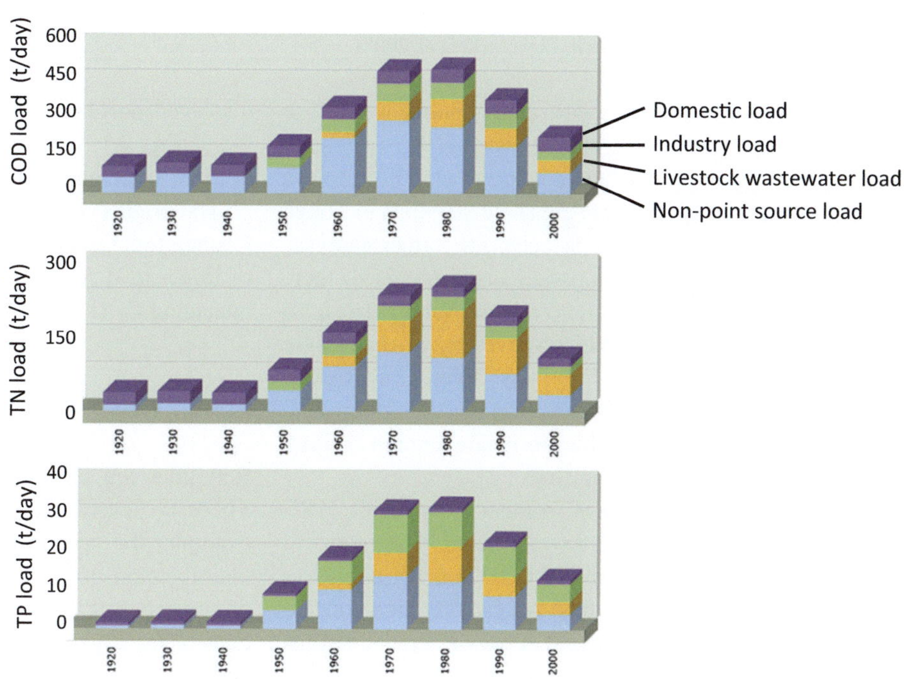

Fig. 2.11 Changes in COD, TN and TP load intensity and fraction into Tokyo Bay from the watershed (Based on http://gis2.meic.go.jp/TokyoGulfWebsystem/)

fish mortality occurs at intervals of a few years, while 4500 fishery parties catch 20,000 tonnes annually (based on Tokyo Bay Environmental Information Center data, http://www.tbeic.go.jp/; Tokyo Bay Renaissance Promotion Council 2013). The occurrence of red tides has gradually decreased from 50 to 30 times a year, and the occurrence of blue tides has stabilised at 3 times a year over the past 10 years. Nevertheless, these incidents suggest that Tokyo Bay is suffering from severe eutrophication.

The land-based organic load (represented by the chemical oxygen demand or COD) and nutrients (TN and TP) in Tokyo Bay peaked in the 1980s but decreased markedly thereafter (Fig. 2.11). The COD load was 100×10^3 kgd^{-1} until 1940, increased in proportion with the economic development of Japan after 1940 and reached 580×10^3 kgd^{-1} in the 1980s. After that, it decreased to 320×10^3 kgd^{-1} in the 2000s. TN and TP also show the same pattern of historical changes. The reasons for the decrease in COD, TN and TP load from the 1980s to the 2000s are (1) decreased loads from non-point sources because of increased sewage treatment and (2) decreased loads from factories because of the enhancement of legislative managements.

Ando et al. (2005) prepared a summary report of nutrient distribution in recent decades. They used a dataset from the

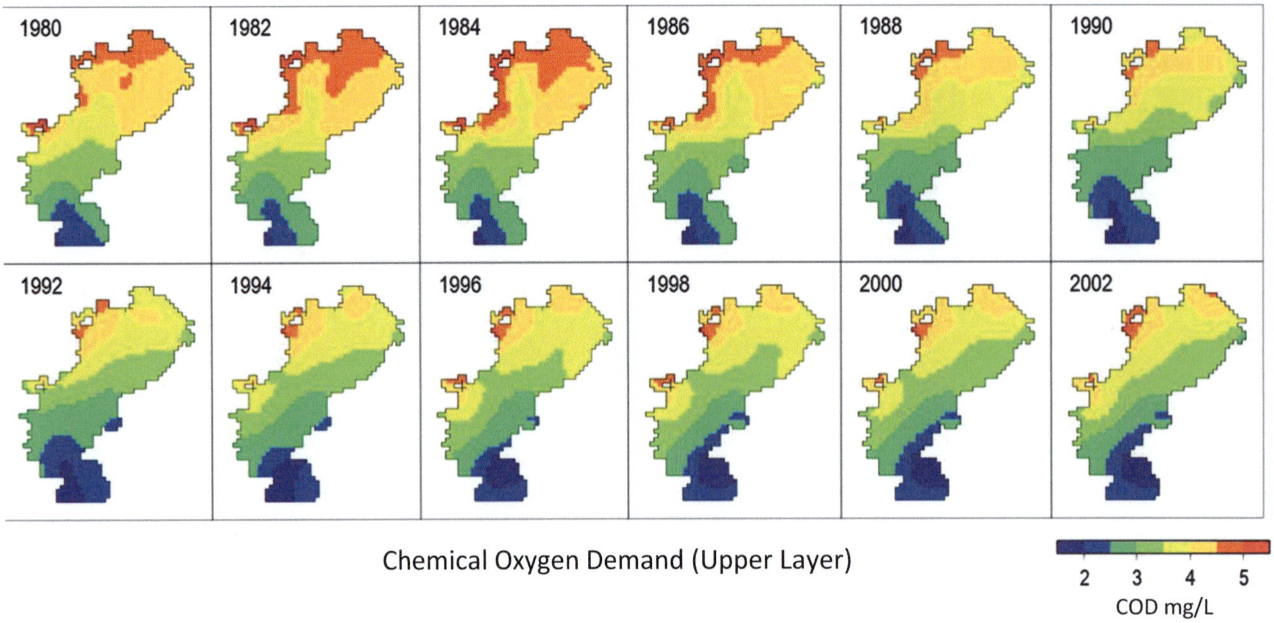

Fig. 2.12 COD distribution in Tokyo Bay (Ando et al. 2005)

1970s for the environmental monitoring of public water. The monitoring was directed by the Ministry of Environment, and sectors of local government conducted monthly surveys and preliminary analysis. There are about 90 points in Tokyo Bay, of which 41 points were selected for further analysis. Because the monthly sampling was unsynchronised, there were fluctuations due to differences in the survey dates. To overcome these fluctuations, observed raw data were decomposed with respect to (a) long-term trends, (b) seasonal changes and (c) short-term noise. Furthermore, spatial interpolation was applied to obtain smooth contour lines for all the areas of Tokyo Bay.

COD is a major component in monitoring. Until 1985, COD was high around the Tokyo Port area. The improvement in water quality in terms of COD concentration was visible near the mouth of the bay. A relatively high concentration of COD extends southward at the western side of the bay, whereas a relatively low concentration of COD extends northward at the eastern side (Fig. 2.12).

TN is an indicator of high levels of anthropogenic activity. Because sewage treatment discharges are densely located in Tokyo Metropolitan City, TN around the northwestern side of the bay has always been high. The worst case was observed in 1988; improvements in COD can be observed near the eastern side of the mouth (Fig. 2.13).

TP has the same contribution from sewage treatment discharge as TN along with significant additional contributions from bottom sediment. This implies that there is an additional high-concentration region in the top of the bay (Fig. 2.14). Nevertheless, the eastern area of Tokyo Bay is now oligotrophic, especially with respect to dissolved inorganic phosphorus (DIP; Hasegawa and Hayashi 2009: Fig. 2.15). *Nori* mariculture has been facing severe nutrient shortage since 1995.

The origins of COD are summarised in Fig. 2.16 (Central Council fora Environment 2004). Only 27 % of COD consists of direct load from land, which explains the difficulties in managing COD with only land-based measures.

2.3.3 Red Tide

Red tides are caused by extreme phytoplankton blooms. There are several criteria for identification of red tides. For example, the criteria of Tokyo Metropolitan City for the identification of red tide condition are (1) less than 1.5 m of transparency, (2) at least 10^3 cells ml^{-1} of phytoplankton density for large species or 10^4 cells ml^{-1} for small species and (3) at least 50 μg l^{-1} of chlorophyll a. Nomura (1998) suggested 30 μg l^{-1} of chlorophyll a as a criterion.

Red tides were an uncommon phenomenon between 1900 and 1910 in Tokyo Bay (Okamura 1907; Asakura 1907); however, when they did occur, the main species of phytoplankton were flagellates such as *Gymnodinium* and *Peridinium* (Nomura 1998). There are few data available regarding red tide occurrences and constituent species prior to 1950. Nomura (1998) presented a historical review of red tide phytoplankton species and occurrences in Tokyo Bay. Since the 1950s, phytoplankton present during red tide conditions have diversified. Red tide conditions with diatoms such as *Skeletonema costatum* began to occur (Nomura 1998). After the 1970s, red tide conditions with

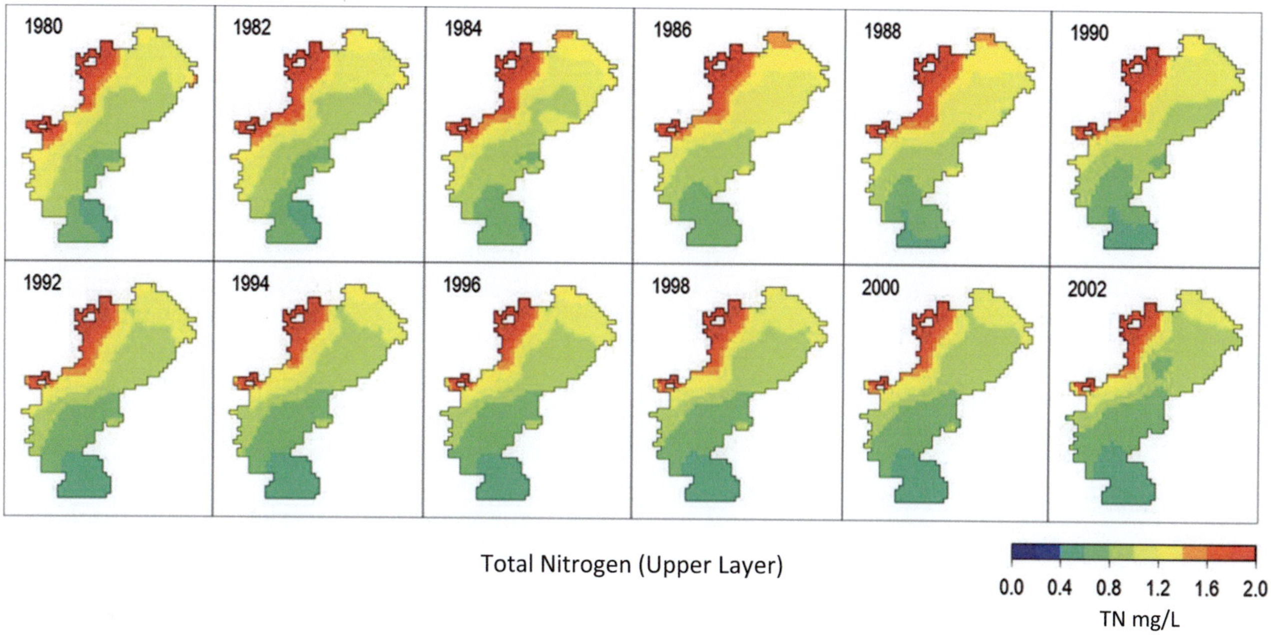

Fig. 2.13 TN distribution in Tokyo Bay (Ando et al. 2005)

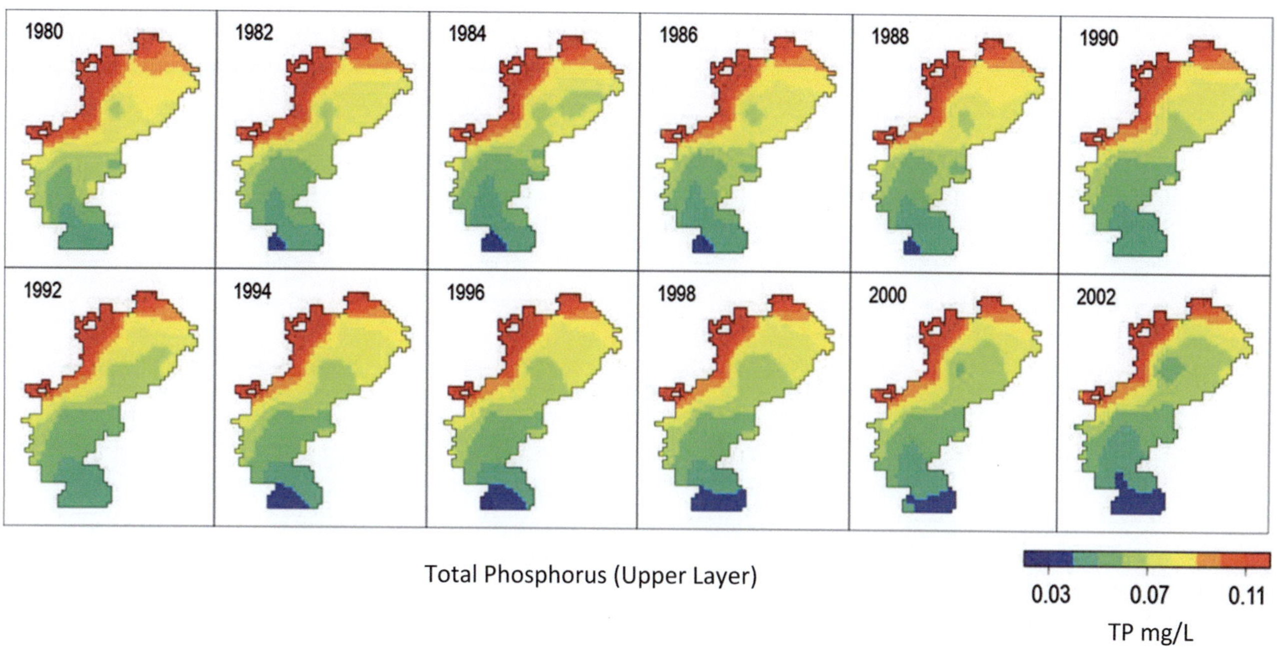

Fig. 2.14 TP in Tokyo Bay (Ando et al. 2005)

S. costatum appeared throughout the year (Furota 1980) and became frequent after the 1980s. Very highly concentrated blooms, however, were composed of flagellates instead of diatom species such as *Heterosigma akashiwo* and *Prorocentrum* (Nomura 1998).

There is no official historical record regarding the frequency of red tide occurrences in Tokyo Bay. Nomura (1998) reported summarised numbers of red tide occurrences from 1907 to 1995 using scientific papers and local government records. There were fewer than 5 days per year before the 1940s. The number gradually increased from the 1950s up to the 1980s, reaching 20 days per year during the 1980s. The frequency has remained constant at 15–20 days per year since then (Tokyo Bay Marine Environment Research Committee 2011). Because red tide criteria and areal coverage differ, the absolute numbers do not match the data shown in Fig. 2.10. Nevertheless, the gradual increase from the 1950s and saturation of high-

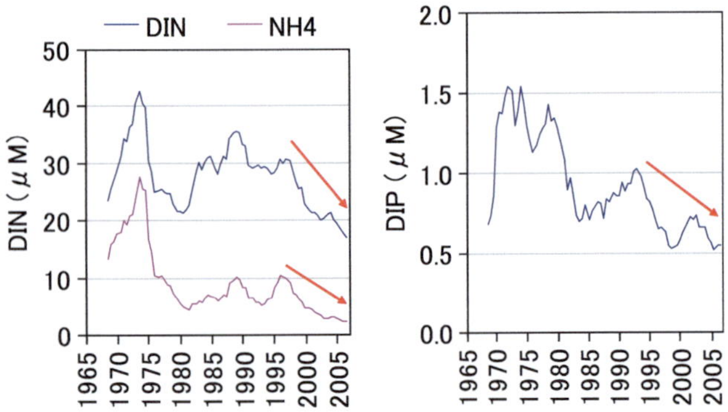

Fig. 2.15 Changes in inorganic components of nutrients in the eastern area of Tokyo Bay (After Hasegawa and Hayashi 2009)

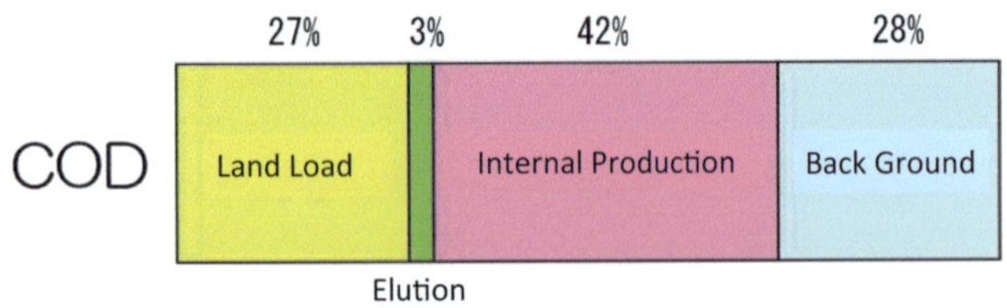

Fig. 2.16 Sources of COD (Central Council for Environment 2004)

level occurrence during the 1980s and 1990s were followed by a gradual reduction in the number of red tide occurrences.

2.3.4 Hypoxia/Blue Tide

The duration of hypoxia in the bottom layer of Tokyo Bay has remained almost constant at 3–4 months per year since the 1980s. The area in which anoxic water is present usually extends over 50 % of the inner bay in summer (July–Sep. Fig. 2.17).

The anoxic water conditions in the bottom layer enhance hydrogen sulphide release. When the anoxic water with hydrogen sulphide upwells to the surface layer, the oxygen in the surface layer oxidises some of the hydrogen sulphide to sulphur, which changes the colour of the surface layer to blue white. This is called a blue tide (aoshio). Hydrogen sulphide is harmful to marine organisms. In Tokyo Bay, blue tide conditions occur in the inner bay when strong north winds continue for several days during summer and autumn. The blue tides inflict significant damage on benthos such as shellfish that cannot escape from the blue tide water.

Blue tide conditions in Tokyo Bay were first observed in the 1950s, as shown in Fig. 2.10. The frequency of blue tides reached its peak in the 1980s at about 6 times per year. Since

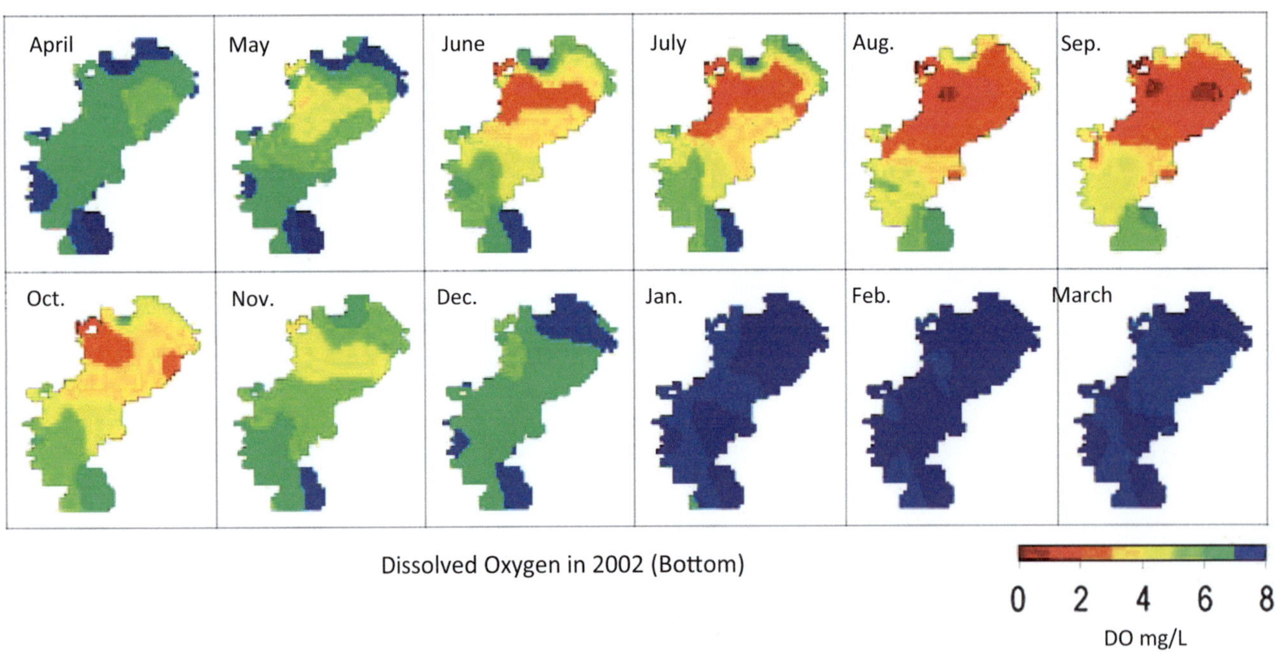

Fig. 2.17 Seasonal development of hypoxia (Ando et al. 2005)

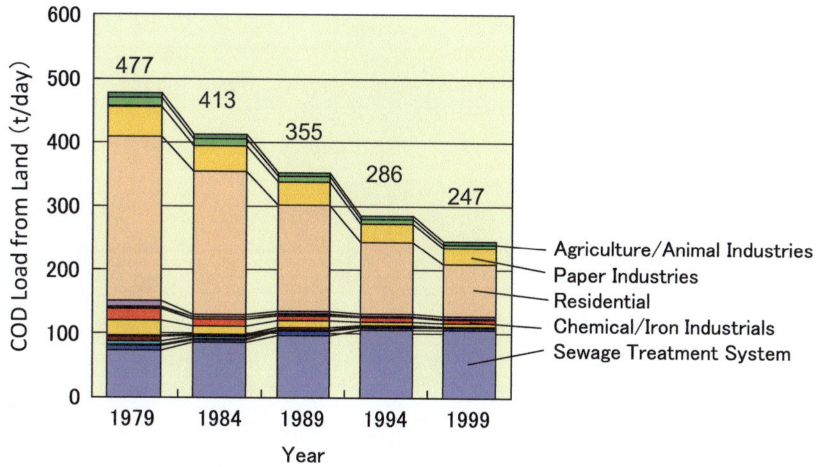

Fig. 2.18 Details of COD load fraction into Tokyo Bay between 1979 and 1999 (Based on Ministry of Environment 2010)

then, it has decreased to the current level of approximately 3 times per year, with each event persisting for 2–3 days on average.

2.3.5 Tidal Flats and Benthos

The area of tidal flats has decreased by approximately 90 % over the last 100 years (Ogura 1993). An internal zone benthic biomass of 12.6×10^4 tonnes was lost as 126 km^2 of tidal flats was reclaimed (Furota 2005). Furthermore, Kakino (1986) suggested that 3×10^4 tonnes of bivalve biomass were killed by blue tide conditions in September 1985, which demonstrates that the richness of biological production in the shallows and tidal flats is under the threat of disappearance.

The decrease in benthic biomass has resulted in a decrease in the purification capability of Tokyo Bay (Furota 2005), which is considered to be the main reason why the frequencies of red tides and hypoxia in Tokyo Bay have not decreased, even though the COD, TN and TP loads have decreased. Without sufficient benthic organisms, material circulation cannot be healthy, as discussed in Sect. 2.3.5.

2.4 Responses

2.4.1 Effect of Load Control

After 1971, when the Water Pollution Control Act was enacted, significant reduction of COD input loads was achieved (Fig. 2.18). In particular, the levels from household and industrial waste water decreased by half during 1979–1999. Furthermore, because of the rapid implementation of sewage systems, an increasing load was reported from the water treatment system.

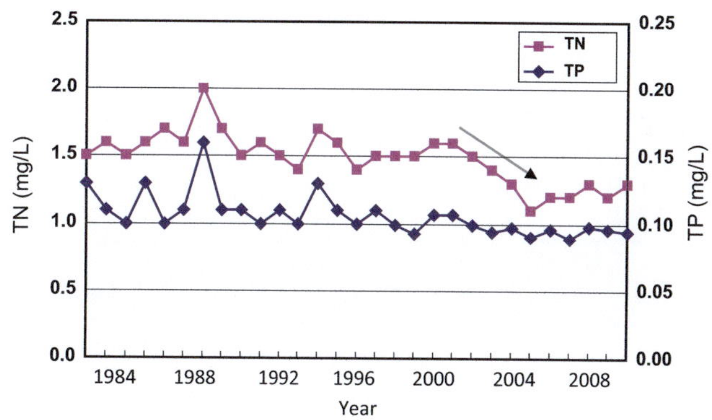

Fig. 2.19 Changes in spatially averaged TN and TP on the northwestern side of Tokyo Bay (http://www.kankyo.metro.tokyo.jp/water/tokyo_bay/measurements/index.html) (Tokyo Metropolitan City 2011)

The results of these efforts are now visible in bay-wide distribution and local trends in COD, TN and TP. As illustrated in Fig. 2.12, the COD high-concentration area in the inner part of the bay has significantly reduced. In Figs. 2.13 and 2.14, TN and TP show the same trend of decrease in the high-concentration area since the 1980s; however, this is not significantly visible at the bay-wide scale. Figure 2.19 shows a detailed view of TN and TP for the northwestern Tokyo Bay (Tokyo Metropolitan City 2011) and illustrates a significant decrease in TN after 2000. This is considered to be a result of the implementation of sewage treatment systems (almost 100 % installation was completed in 1995). Notwithstanding, TP appears to have undergone a more moderate change because the contribution from sediment-based circulation is thought to have compensated for the effects of decrease in input.

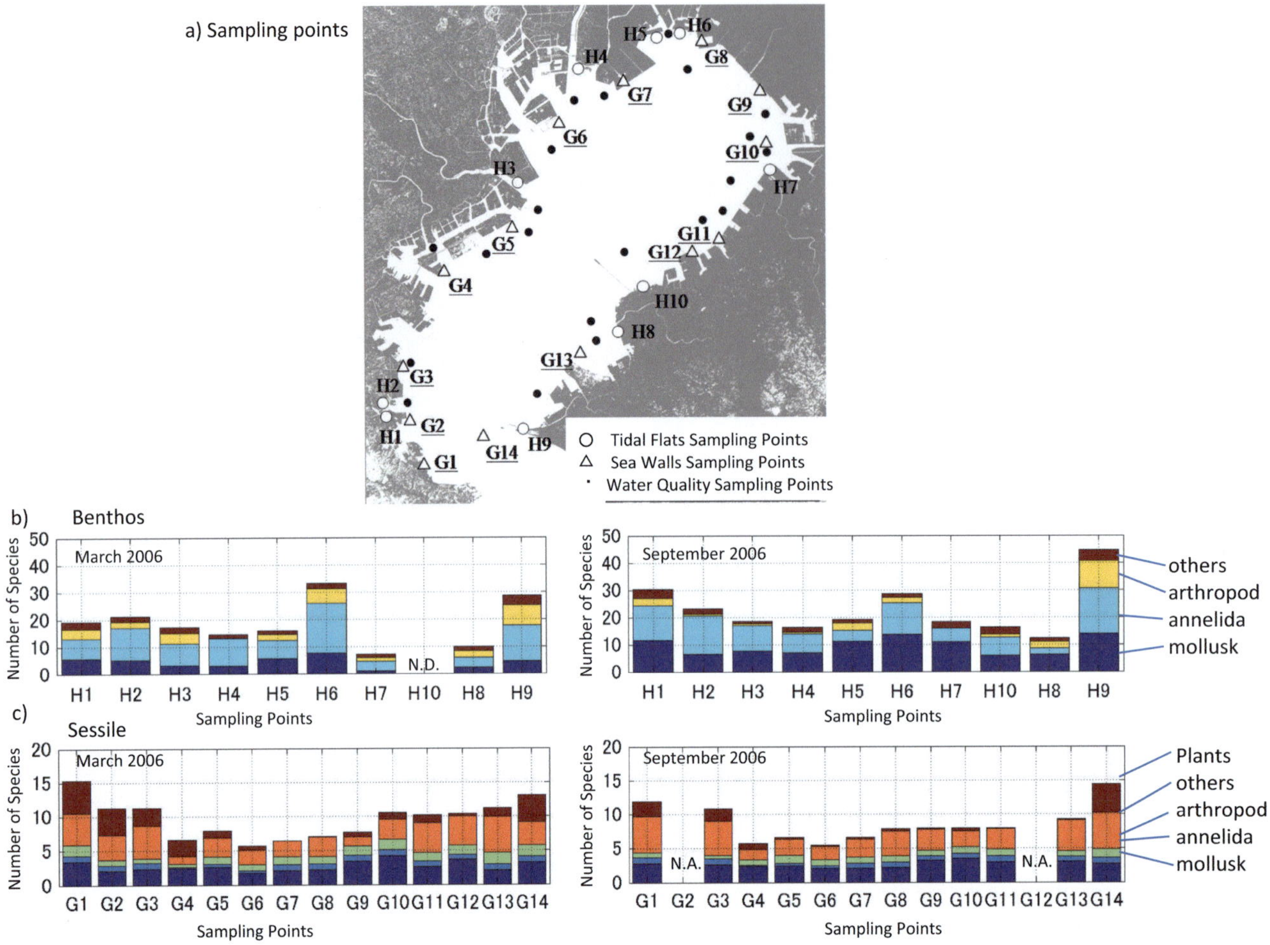

Fig. 2.20 Distribution of sessile and benthic organisms in Tokyo Bay (After Igarashi and Furukawa 2007)

2.4.2 Potential of the Bay as a Habitat

The environmental changes in Tokyo Bay could have affected the distribution of living organisms in the bay. Igarashi and Furukawa (2007) drew a distribution map of benthos on tidal flats and sessile organisms on seawalls. The data showed that: (1) even in the inner part of the bay in a highly eutrophic area, the tidal flats maintained a rich benthic diversity while almost no benthos was recorded in the northern middle part of the bay and (2) the diversity of sessile organisms was attributed to changes in the water quality such as transparency and salinity. Anthropogenic impact is one of the possible causes of these distributions. The inner part of Tokyo Bay suffers from severe hypoxia in summer and autumn (Fig. 2.20).

Field observation of the sessile assemblages on coastal structures was conducted along the Japanese coastline, including three bays (Tokyo Bay, Ise–Mikawa Bay and Osaka Bay) and five other main areas (Tomakomai, Akita, Niigata, Maizuru and Dokai Bay) (Table 2.2: Kamimura et al. 2011). *Mytilus galloprovincialis* was the primary dominant species in Tokyo Bay, Akita and Niigata, and other bivalves, crustaceans and polychaetes were the primary dominant species in other bays and areas. The primary dominant species in Dokai Bay has changed dramatically over the past 20–30 years, as reported in Sect. 4.3.2. In the three bays, the species diversity of the assemblage was lowest at the inner side of the bays, where the primary dominant species was the most dominant. Although the peaks in species diversity were localised in specific parts of Tokyo and Osaka Bays, diversity was distributed randomly in the Ise–Mikawa Bay. Data from the five areas indicate that species diversity declines as latitude increases. It was not clear whether coastal structures and material affect sessile species diversity. Species abundance and diversity were not correlated with the number of years after construction, which implies that the sessile assemblages on vertical coastal structures are established immediately after construction and succession occurred over a short period. The maximum number of species decreased with increasing wave height, as indicated in previous studies (Asai et al. 1997). The maximum number of species appears between 1.0 and 6.0 m below the mean sea level, and the abundance and wet weight of each population were highest

Table 2.2 Abundance of sessile organisms in bays (after Kamimura et al. 2011)

Place	Tokyo Bay	Ise–Mikawa Bay	Osaka Bay	Dokai Bay
Number of sampling points	15	15	15	1
Number of species in the bay/area	159	185	264	114
Species diversity in sampling point	44.5	48.7	96.9	114
Dominant species	*Mytilus galloprovincialis*	*Xenostrobus securis*	Ophiactidae	*Cirriformia* sp.
	Chthamalus challengeri	*Chthamalus challengeri*	*Chthamalus challengeri*	*Xenostrobus securis*
	Ophiactidae	*Melita* sp.	*Mytilus galloprovincialis*	
	Thais clavigera	*Elasmopus japonicus*	*Balanus amphitrite*	
		Mytilus galloprovincialis	*Balanus trigonus*	
		Amphibalanus improvisus	Serpulidae	
		Hyale sp.		
		Ophiactidae		
		Nanosesarma gordoni		

at the mean sea level. Species diversity increased with increasing depth of the sampled layer. Around mean sea level, the abundance of certain species increased, which reduced species diversity.

2.4.3 Ecosystem-Based Management

Another important aspect is the distribution patterns of the species in the bay. Several living organisms have unique life stages. Larvae, juveniles and adults are transported and live in different places at different stages of their life cycle, forming an interstage larval network. This network is established with nodes and links. Nodes are the habitats, and links are the transport systems in the bay, which are mainly facilitated by water circulation. Because anthropogenic impacts affect the habitats and transportation of various organisms, urbanisation could destroy the functional integrity of such ecological networks. Hinata and Furukawa (2006) illustrated this phenomenon using the short-necked clam (*asari*) larvae network in Tokyo Bay (Fig. 2.21). Their study demonstrated the existence of an ecological network. Moreover, it showed that the ecosystems in Tokyo Bay could be improved if suitable spawning grounds, even if small, were to be conserved and restored. The ecological network has been considerably weakened, especially in the northern to western part of the bay due to changes in the hydrology.

2.4.4 Tokyo Bay Renaissance Project

On 26 March 2003, the Tokyo Bay Renaissance Promotion Council, which is composed of six central government agencies[2] and eight regional government bodies,[3] endorsed the 'Action Plan for Tokyo Bay Renaissance' (Fig. 2.22). The goal of the action plan is to restore the beautiful coastal environment for people to enjoy and to sustain its natural biodiversity. This is to be achieved within 10 years through collaboration among the agencies involved (by 2012) and is subject to an annual review of related activities. Two internal appraisals were conducted in 2006 and 2009, and management adopted their findings.

The action plan was initiated as a part of the Urban Renaissance Project of the Japanese Cabinet (Tokyo Bay Renaissance Promotion Council 2003). The council was set up with three working groups, namely, WG1 on land-based implementation, WG2 on sea-based implementation and WG3 on monitoring.

The process for assessing the achievement of the goal is unique. Several monitoring sites with specific targets were defined in an area identified for priority implementation. Each site has a specific target plan for restoration, and thus project assessment in terms of specific goals can be undertaken. The priority implementation area matched the coastline that has a relatively weak ecological network, as discussed in Sect. 2.2.1. Thus, the local enhancement of ecological functions at the target sites will contribute to the holistic restoration of the bay.

[2] Ministry of Land, Infrastructure, Transport and Tourism; Japan Coast Guard; Ministry of Agriculture, Forestry and Fisheries; Forestry Agency; Fisheries Agency; Ministry of the Environment and Cabinet Secretariat.

[3] Tokyo Metropolitan City, Chiba Prefecture, Kanagawa Prefecture, Saitama Prefecture, Yokohama City, Kawasaki City, Chiba City and Saitama City.

Fig. 2.21 Schematic diagram of the *asari* larval network in Tokyo Bay. Numbers indicate the comparative quantities of larvae transported between local regions (After Hinata and Furukawa 2006)

Fig. 2.22 Structure of the Tokyo Bay Renaissance Promotion Council

Several sites to be monitored and assessed for restoration and management (called monitoring points, Fig. 2.23) have been designated in an area of priority implementation. These areas were determined using ecosystem-based management strategies. The larval network of *asari* (short-necked clam) has deteriorated in the upper western side of the bay, and major discharge is occurring in this area.

Furthermore, each monitoring point has specific targets for restoration, allowing the use of objective and quantifiable indicators of progress towards the restoration objectives (such as biodiversity or the dissolved oxygen (DO) level of bottom layer). The expectation is that achieving these appeal point targets will also contribute to enhanced ecosystem health in the surrounding areas. In addition, by providing node habitats, successful activity around the appeal points

Fig. 2.23 Prior implementation area and monitoring points of action plans for Tokyo Bay Renaissance

could strengthen environmental connectivity throughout the bay and hence promote overall bay ecosystem health. Finally, activities at the appeal points helped develop a model for urban-type *sato-umi* implementation. *Sato-umi* is a coastal landscape that is enhanced by its rich biodiversity and high productivity as a result of effective use by humans (Yanagi 2007).

2.4.5 Revision of Action Plan for Tokyo Bay Renaissance and Public–Private Cooperation

After 10 years of implementation of the action plan for the Tokyo Bay Renaissance, the renaissance council reported their accomplishments (Tokyo Bay Renaissance Promotion Council, 2013). The report pointed out the positive effects of projects on land load control, which resulted in improved water quality of the bay, and restoration and construction of tidal flats and shallows, which increased the biomass and richness of benthos in the bay. Nevertheless, the DO level of the bottom layer in summer season has not recovered from the hypoxic environment.

The report highlighted two major challenges that should be addressed as the next step of the Tokyo Bay Renaissance. One of the challenges is an assignment of detailed environmental indicators to assess various related projects under the action plan. The other challenge is expanding active participation in Tokyo Bay Renaissance activities.

The DO level is an indicator for results of material cycling through watershed to bay environment. It is a good indicator for the final assessment of the goal, but is difficult to use for assessing indirect respective efforts or contributions that enhance the environment such as a beach cleanup campaign by a non-profit organisation (NPO) group. The more attractive and understandable indicator is required to promote in-kind contribution from various stakeholders.

The restoration of the Tokyo Bay environment cannot be accomplished in a short period by drastic measures. There is no easy route to restoration. However, it can be achieved by accumulating possible small-scale efforts in a sustainable manner. Adaptive management is also a key concept for sustaining these efforts. To implement these strategies, the participation and cooperation of both the public and the private sector should be enhanced.

In 2013, an action plan for the Tokyo Bay Renaissance (2nd phase) was set by the council (Tokyo Bay Renaissance Promotion Council 2013). The structure of the plan followed that of a previous action plan and also reflected the assessment report. There are three new points in the plan: (1) increased emphasis on targets that include the promotion of the *Edo-mae* culture, (2) suggestions for establishing a new institutional arrangement for enhancing public–private partnerships and (3) addition of various success indicators for assessing the accomplishments of the plan.

A public–private cooperation forum for the Tokyo Bay restoration was established in November 2013 with the participation of central government; local governments; industry; fisheries; the leisure, historical and cultural sectors; various NPOs; and the general public. This forum can form a project team (PT) for the specific task of drafting a proposal for the Tokyo Bay Renaissance Promotion Council. Then, the forum will obtain a consensus from the forum members in a general annual assembly and convey the official proposal to the council. Through this process, the link between public and private sectors will be maintained in a sustainable manner.

2.5 Remaining or Restored Environments in Tokyo Bay

There are many remarkable shorelines in Tokyo Bay. Brief descriptions of remaining or restored environmental assets are given in Fig. 2.24.

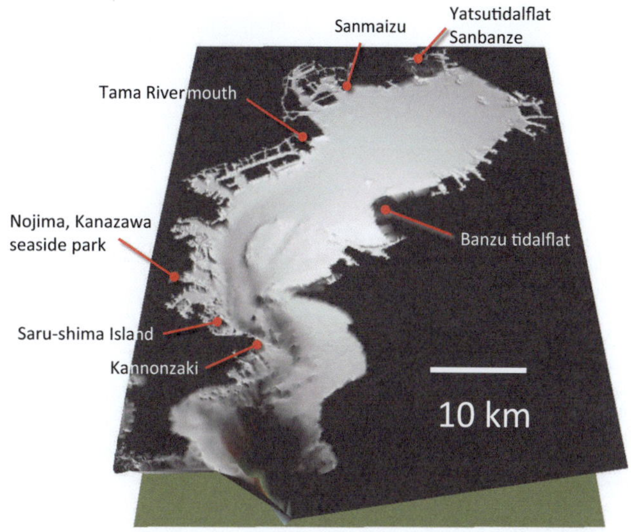

Fig. 2.24 Remaining or restored environmental assets in Tokyo Bay

2.5.1 Chiba Prefecture

2.5.1.1 Obitsu River Mouth and Banzu Tidal Flat

The discharge of the Obitsu River is 620,000 m^3s^{-1} and supplies sediment to the foreshore. The Banzu tidal flat (Fig. 2.25) is the largest remaining tidal flat in Tokyo Bay. The shoreline is approximately 8-km long, and the width of the flat is about 2 km. Massive *nori* mariculture operates during winter, and short-necked clam harvesting is a popular leisure activity during spring and early summer. The *Aqua-Line* (a bridge–tunnel combination across Tokyo Bay) offers good access to the Banzu tidal flat from the Tokyo area via tunnel and bridge. Local fishermen have formed a council to promote fisheries and leisure.

2.5.1.2 Yatsu Tidal Flat and Sanbanze

The Yatsu tidal flat (Fig. 2.26c, d) is a foreshore tidal flat. Due to land reclamation, only 40 ha of tidal flat remains in the well-developed area. More than 110 species of birds have been observed at the site. Because of its importance for migrating birds, the Yatsu tidal flat is registered as a Ramsar site.

Sanbanze (Fig. 2.26a, b) is an enclosed tidal flat with a size of 4 km × 2 km. The mean water depth is less than 2 m, and the flat is well used for *nori* mariculture and clam harvesting by local fishermen. A development plan was proposed previously by the local government, but it was withdrawn because of public demand for the conservation of the area. Now, restoration committees for the public (Sanbanze forum) and experts (specialist meeting) have been established to find appropriate options for the use of Sanbanze.

2.5.2 Tokyo Metropolitan City

2.5.2.1 Sanmaizu

Sanmaizu (Fig. 2.27) is a remaining shallow area of the Ara River (discharge 700,000 m^3s^{-1}) and the original Edo River

Fig. 2.25 Panel (**a**) *nori* mariculture and the *Aqua-Line* bridge, (**b**) tidal flat as a rich field for high biodiversity, (**c**) *sudate* traditional net fishing as a leisure activity and (**d**) shell-collecting device used by local fishermen

(discharge 660,000 m^3s^{-1}) mouth delta. Tokyo Metropolitan City constructed a seaside park on the western and eastern sides of the shallow area. The seaside park project was carried out from 1980 to 1989. Before project execution, on-site surveys for natural resources, sediment stabilisation and water quality control were conducted from 1972 to 1984. On the basis of the surveys, basic planning and monitoring were conducted. The total project area extended over 41,150,000 m^2, including a west shore island and an east shore island. The west shore island was planned as a bird habitat that will imitate a natural tidal flat. Public access is limited, and the d_{50} of the sediment is 0.02 mm. The east shore island was planned as a recreational sandy beach open to the public. The geological setting (cross section) was determined by substrate stabilisation against design wave height (PIANC 2003).

2.5.2.2 Odaiba Seaside Park

Odaiba (Fig. 2.28) is a historical reclaimed land that was used as a cannon base during the *Edo* era. Subsequently, because the area is surrounded by breakwaters and reclaimed land, the park has a calm and safe water body measuring 44 ha. There are piers for commuter ferries between other sightseeing spots in Tokyo such as Asakusa. At night, many Yakata boats (traditional Japanese-style cruise boats) can be seen on the pond. Furthermore, the Odaiba Environmental Education Promotion Committee provides a great opportunity for primary school students to experience *nori* mariculture at the park.

2.5.2.3 Tama River Mouth

The Tama River (discharge 320,000 m^3s^{-1}) has one of the largest riverine tidal flats in Tokyo Bay (Fig. 2.29). The tidal

Fig. 2.26 Panel (**a**, **b**) *nori* mariculture, birds and shell collection at Sanbanze tidal flat, (**c**) nature observatory centre at the Yatsu tidal flat (from the Narashino City web site; http://www.yatsuhigata.jp/) and (**d**) scenery of the Yatsu tidal flat from the observatory

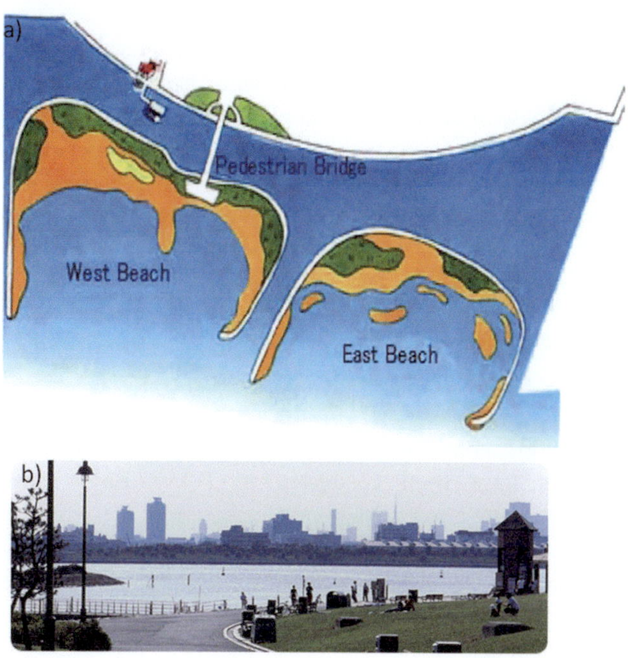

Fig. 2.27 Panel (**a**) schematic image of the seaside bird park at Sanmaizu and (**b**) view of the park

flat area is 45 ha, and rich reed fields are established. Seawater intrusion forms a salt wedge in the river mouth. In summer, the seawater is often hypoxic. Thus, the abundance of fish is highly dependent on the salt wedge occurrence. This area is also popular as a public observatory. Birdwatching, plant observation, benthos monitoring and fish censuses are often organised.

2.5.3 Kanagawa Prefecture

2.5.3.1 Nojima and Kanazawa Seaside Park

Nojima (Fig. 2.30a, c) is the last remaining natural shoreline on the Tokyo Bay side of Kanagawa Prefecture and has been used as a calm sandy beach. Subsequently, land reclamation has broken up the shoreline and deterioration of seagrass meadows has occurred. A consortium of NPO groups promotes eel grass transplantation in both Nojima and the Kanazawa seaside parks.

The Kanazawa seaside park (Fig. 2.30 b, d) has a constructed shoreline length of about 1 km and a width of about 200 m. It was constructed during 1978–1979 using

Fig. 2.28 Panel (**a**) schematic image of Odaiba seaside park, (**b**) cannon base of the *Edo* era, (**c**) *nori* mariculture as environmental education and (**d–e**) *nori* harvesting by students

sand from a mountain that was placed under seawater for 5 years before use. From April to June, the park is mainly used for shell-collecting activities, with a total of 600,000 visitors (40,000 people/day in peak season). From July to August, the park is mainly used for sea bathing with a total of 700,000 visitors (30,000 people/day in peak season).

The area is highly productive in terms of fisheries. Nevertheless, during early summer, drifts of Ulvales (Ulvaceae) form on the beach, and cleaning operations are carried out by Yokohama City. In total, about 800 t of Ulvales is collected annually.

2.5.3.2 Kannonzaki and Saru-shima Island

Kannonzaki (Fig. 2.31 a, b) is a part of a prefectural park (67 ha). This peninsula is located in the narrowest channel of Tokyo Bay with Cape Futtsu. The width of the bay narrows to about 7 km at this section. The depth of the channel is more than 50 m, and the fastest tidal current exceeds 1 m s^{-1}. There are small pocket beaches and coastal cliffs that form picturesque spots.

Saru-shima (Fig. 2.31 c, d) is a small island and the only natural island in Tokyo Bay. Saru-shima had been used as a cannon base from the Edo era until the Meiji era. Since 1995, Yokosuka City has managed the site and has operated a commuter ferry. At present, the island is a part of the eco-museum of Yokosuka City.

2.6 Future Tasks

2.6.1 Monitoring Project

Since 2008, annual monitoring campaigns of the bay environment have been conducted in Tokyo Bay. These campaigns are a part of the implementation of the activities of working group 3 of the Action Plan for Tokyo Bay Renaissance. Monitoring was performed both in the sea and in the watershed to obtain a holistic view. During the 2008 campaign, data were corrected by 46 institutions at 568 monitoring points. The number of monitoring points

Fig. 2.29 Panel (**a**) rich reed field of a tidal flat; (**b**) benthos monitoring shows a unique estuarine environment and (**c**) the Tama River is famous as a fishing spot

increased up to 820 in 2011. DO, salinity and temperature were monitored in sea area from the surface to the bottom with 1-m vertical spacing between the monitoring points. COD, temperature and discharge were monitored in river areas. These datasets provide a detailed view of hypoxic water occurrences in the inner part of the bay. Furthermore, a conceptual model for how DO distributions have changed and transported in the bay can also be constructed. For such interpretation, surface circulation data monitored by oceanographic radar systems, continual water quality data monitored by monitoring posts and some knowledge of water circulation mechanisms in the bay is required. One trial of the campaign is not sufficient to answer all the questions; nevertheless, this trial demonstrates the importance and efficiency of monitoring and management of the bay environment.

After the monitoring, all participants are invited to participate in follow-up workshops to compile the results into a Tokyo Bay Environmental Map (Fig. 2.32). These workshops began as a voluntary session organised by the National Institute for Land and Infrastructure Management. At present, workshops are identified as an official component of interpretation and enhancement in the monitoring campaign. Outputs from the workshop are incorporated into the map, and the results of discussion on the improvement of monitoring efficiency and quality are reflected in the next monitoring plan.

Adaptive improvements of the monitoring scheme and sustaining of the activity are important future tasks for the understanding of Tokyo Bay.

New items to be monitored are discussed in the Ministry of Environment. The ministry produced a medium- to long-term vision for enclosed seas in 2010 (Ministry of Environment 2010). The necessity for environmental evaluation of habitat suitability, including not just eutrophication indicators, such as COD, TN and TP, but also additional indicators, such as transparency and DO, was discussed. Some long-term scenarios for environmental management were tested. Only little changes are predicted for DO, while drastic improvement in transparency and COD and moderate improvement in TN and TP are predicted for Tokyo Bay, Ise Bay and Seto Inland Sea (Ministry of Environment 2010; Fig. 2.33).

In 2011, the National Institute for Land and Infrastructure Management (NILIM) conducted bay-wide monitoring of

Fig. 2.30 Panel (**a**) remaining natural shoreline at Nojima, (**b**) constructed seaside park at Kanazawa (the dark colour underwater is restored eel grass meadows), (**c**) eel grass plantation project at Nojima and (**d**) one of the most crowded locations at Kanazawa seaside park

transparency (Furukawa and Ishimaru 2012). As shown in Fig. 2.34, 273 points were monitored on the same day; the distribution of transparency with temporal changes is plotted in Fig. 2.35. Because transparency is correlated with suspended solid concentration (SSC), phytoplankton and COD, as shown in Fig. 2.36, the distribution appears to be affected by the resuspension of bottom sediment by waves in shallow water, plankton aggregation caused by solar irradiation and run-off from rivers. These results show the potential of transparency as a good indicator for ongoing environmental processes in the bay. Nevertheless, because of such variability in transparency, the monitoring scheme should be carefully devised.

2.6.2 Habitat Creation

Selected implementation of habitat creation was promoted through working group 2 of the Action Plan for Tokyo Bay Renaissance. As a candidate for a new target site to facilitate the balance between economic use and environmental conservation, a habitat creation project was implemented at Shibaura Island, Tokyo, in December 2005 (Sakurai et al. 2008). Shibaura Island faces Shibaura Canal, which is connected to the bay through two sluices and receives major sewage inflows. Therefore, it is a typical urban brackish-water area with a presumably complex flow pattern. The experimental facility for habitat creation in the canal has two 4-m × 8-m pools (pools A and B), which are 0.5-m deep with a sandy bottom on a rocky terraced seawall. The experimental facility at Shibaura Island was situated at a lower height than the high water level but higher than mean water level. The depth of the pool was set at 0.5 m below the surrounding surface. Each pool has a small inlet section (0.1-m deep, 1.5-m wide) through the seawall to enhance water circulation during inundation and drainage (Fig. 2.37). Partnership-type collaboration was employed to encourage active citizen participation in the project, which is supported by scientists and local governments. Specially

Fig. 2.31 Panel (**a**) Kannonzaki park, (**b**) boardwalk around Kannonzaki park, (**c**) Saru-shima Island and (**d**) historical fort in Saru-shima Island (from the Yokosuka City web site; http://www.city.yokosuka.kanagawa.jp/4130/sisetu/fc00000431.html)

designed monitoring methods were used by the participants in the project. The potential of the facility as a suitable habitat for fish and benthos was proven to be high.

Tidal courses (channels, creeks or gullies) are distinctive and important features of natural coastal environments. In particular, tidal-influenced courses have distinctive circulation features that drain freshwater to the sea and inundate tidal flats, mangroves and salt marshes with seawater (Perillo 2009). Okubo (1973) pointed out that lateral embayments along a course enhance longitudinal mixing. Thus, habitat restoration in tidal courses should consider this type of circulation features.

Satoh et al. (2006) illustrated the complexity of water circulation in the canal. The secondary canal was stratified and water in the primary canal was well mixed. Nevertheless, these differences were not simply determined by the canal being secondary or primary but instead by freshwater input, tidal range and topography. Detailed water circulation was determined by float experiments that incorporated public participation. These experiments showed density-driven circulation (estuary circulation) due to stratification in limited places. This circulation introduced seawater to the bottom layer of the canal. Thus, it is assumed that the canal network acts as a tidal course, but in a limited area.

Six months after the construction of the facilities, juvenile gobies (*Acanthogobius flavimanus*), mullets and sandworms populated the pools (Fig. 2.38). The mean size of the gobies in the pools and the surrounding water indicated a significant role of pools. All gobies in the pools were sampled by draining, and their lengths were measured by the image processing of photographs. It appeared that deeper water can serve as habitat for the larger gobies. This is presumably correlated with the hydrology of the pool (shallower water has warmer water in summer, which is rich in DO).

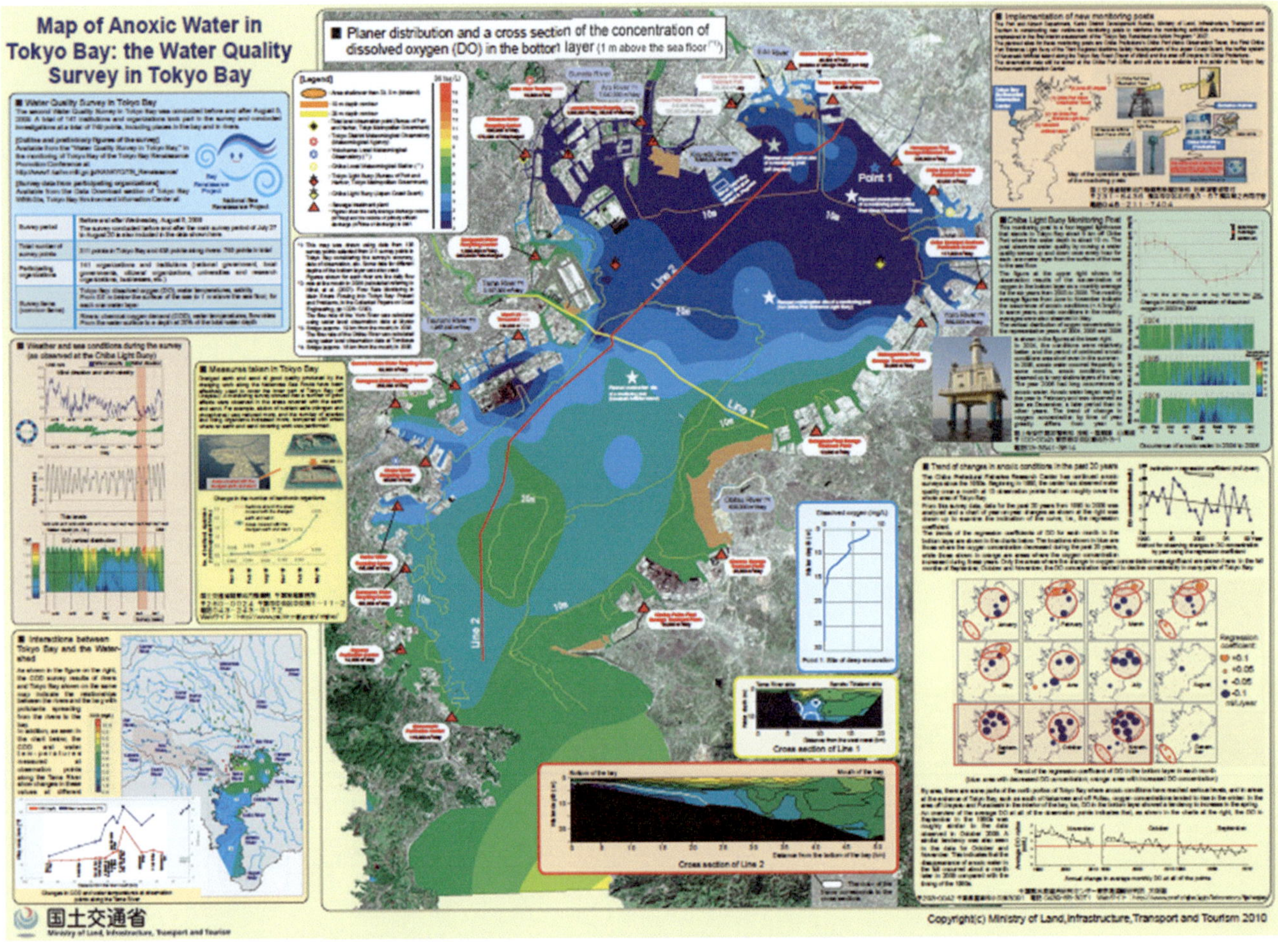

Fig. 2.32 Monitoring Map of Tokyo Bay (Ministry of Land, Infrastructure, Transport and Tourism)

Fig. 2.33 Predicted environmental changes during long-term scenarios (ME 2011)

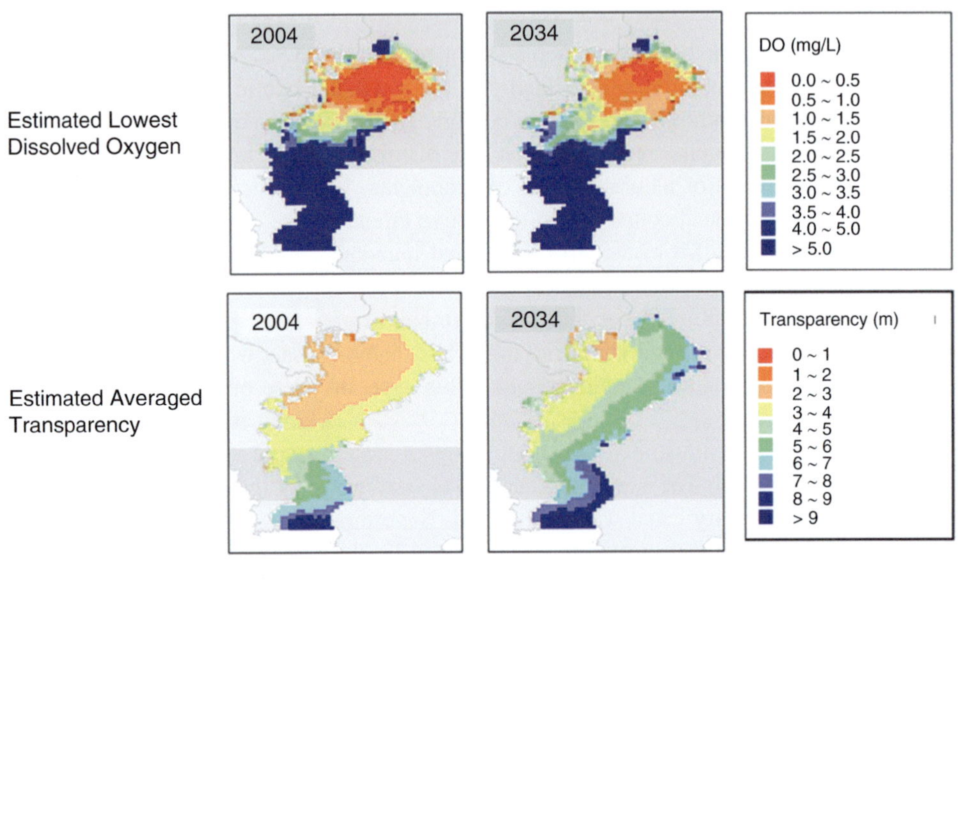

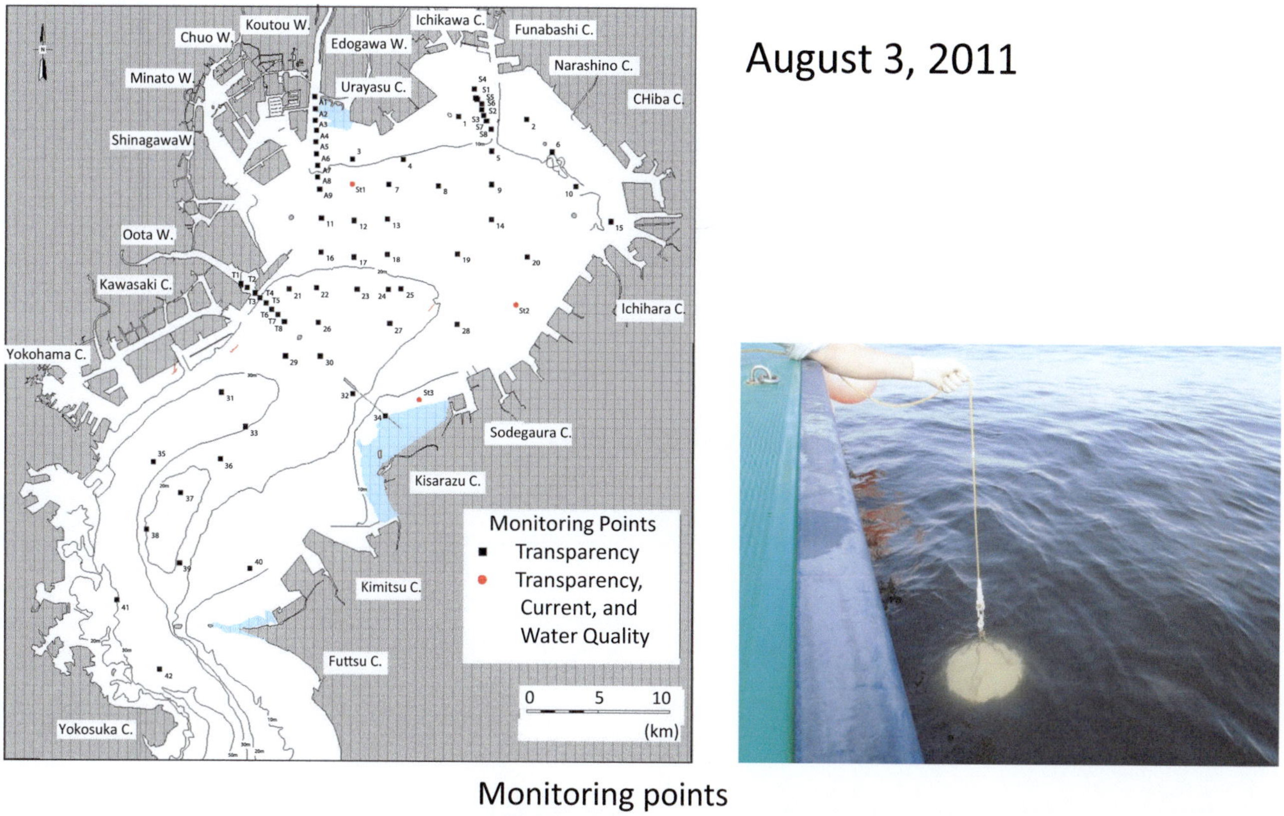

Fig. 2.34 Monitoring points and a snapshot of transparency observation (Furukawa and Ishimaru 2012)

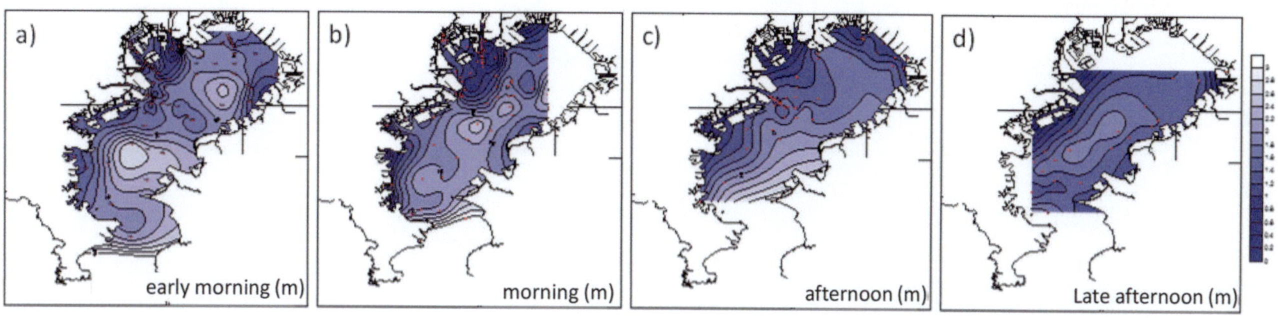

Fig. 2.35 Temporal changes in the spatial distribution of transparency on 3 August 2011 (From *left*: before 10:00, 10:00–12:00, 12:00–15:00 and after 15:00)

These findings show that the constructed habitats (pools) do not act mainly as an ad hoc habitat for gobies but are surrounded by dynamic estuarine system and used as habitat to support some life stages of gobies. In other words, habitat construction appeared to enhance the ecological network in the canal–bay environment.

With good understanding of water and material circulation in canals, it is important to enhance its environmental services by conserving and restoring habitats. This is an important future work.

2.6.3 Public Participation and Adaptive Management

The soft approach to urban wetland management involved stakeholders from NPO, government and researchers. Such organisations have their own objectives. Thus, the target is already set before the project starts, and the project is method oriented. To achieve the target, it is necessary to implement appropriate adjustment based on the

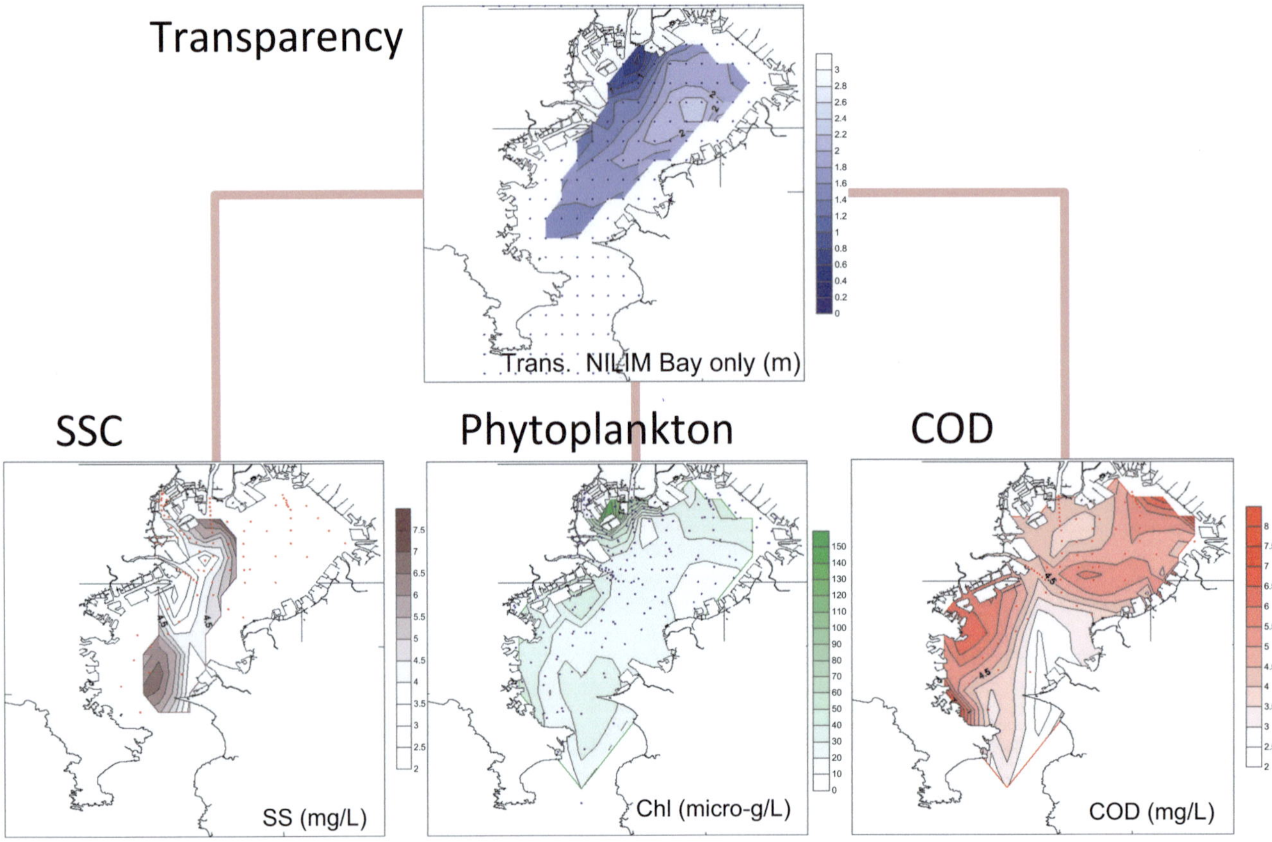

Fig. 2.36 Schematic image of correlation between transparency and water quality

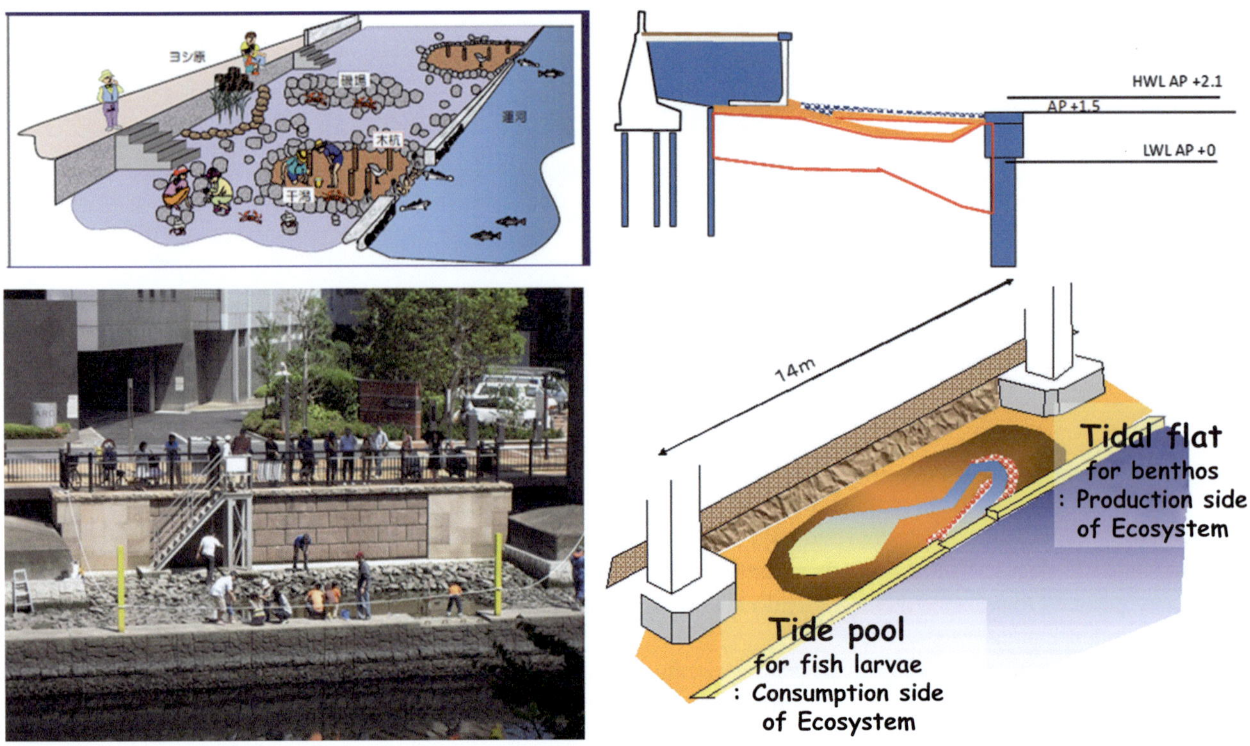

Fig. 2.37 Shibaura Island project site and conceptual diagram of the pool

understanding of the situation with regular monitoring (adaptive management; PIANC 2003).

The Yokohama Port and Airport Technology Investigation Office, Ministry of Land, Infrastructure, Transport and Tourism (MLIT), constructed a terrace-type tidal flat along the existing seawall (Oomura et al. 2009, Fig. 2.39). Teams from the private and public sectors were invited to monitor and maintain the tidal flat. After a year, the construction, structure and morphological stability were recorded. The rapid rehabilitation process by abundant benthic animals was also monitored.

For the first 6 months after construction, colonisation by benthic animals was very rapid, especially for the short-necked clam. The number of short-necked clams increased in the middle terrace with a maximum density of about 20,000 in. m^{-2}, which is the highest density observed in the richest habitat in the region.

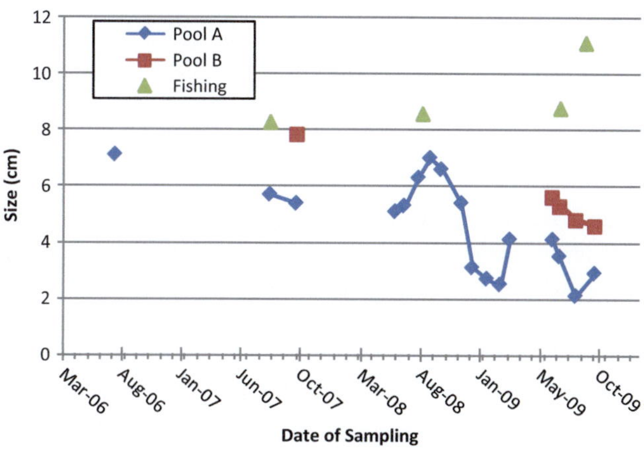

Fig. 2.38 Mean size of sampled gobies in pools A and B and the surrounding water. In pools A and B, censuses were performed by draining the pools with a pump and counting all the fish trapped in each pool. The gobies in the surrounding water were caught by fishing for 2 h. Both sampling events were conducted as public participation events as part of the project (Furukawa 2011)

However, in September 2009, the tidal flat suffered from a massive kill of bivalves. One of the possible explanations was the inflow of anoxic bottom water from the bay. This seemed unlikely because a shallow area like the tidal flats is usually a DO-rich environment. It is difficult to believe that only one short-term intrusion of anoxic water could result in such massive bivalve mortality.

The monitoring record for DO in the middle terrace showed DO recovery in September during daytime inundation, while DO reduction occurred in early summer (July) during night-time inundation (Fig. 2.40, left). The reduction in DO was presumably caused by local consumption of oxygen by benthos respiration. This can occur through a combination of diurnal solar radiation and a semidiurnal tide. In late summer (September), this causes many occurrences of anoxic water, even in the shallow areas (Fig. 2.40, right). It is possible that the benthos was already damaged by the local hypoxic environment when the anoxic water intrusion triggered the massive mortality event. This hypothesis, however, is not confirmed and more monitoring and observations are required to understand this phenomenon.

The monitoring group conducted monitoring and participated in management practices for the establishment of an attractive benthos habitat. Despite their efforts, the massive deaths of benthos could not be prevented. The group is continuously searching for more effective ways to manage the tidal flat with the help of scientists and the public sector.

2.6.4 Public Participation and Adaptive Management 2

Another example of public participation through fishing activities was performed as a new item in the monitoring campaign of Tokyo Bay.

Fig. 2.39 Terrace-type tidal flat constructed at the Yokohama Port and Airport Technology Investigation Office (After Oomura et al. 2009)

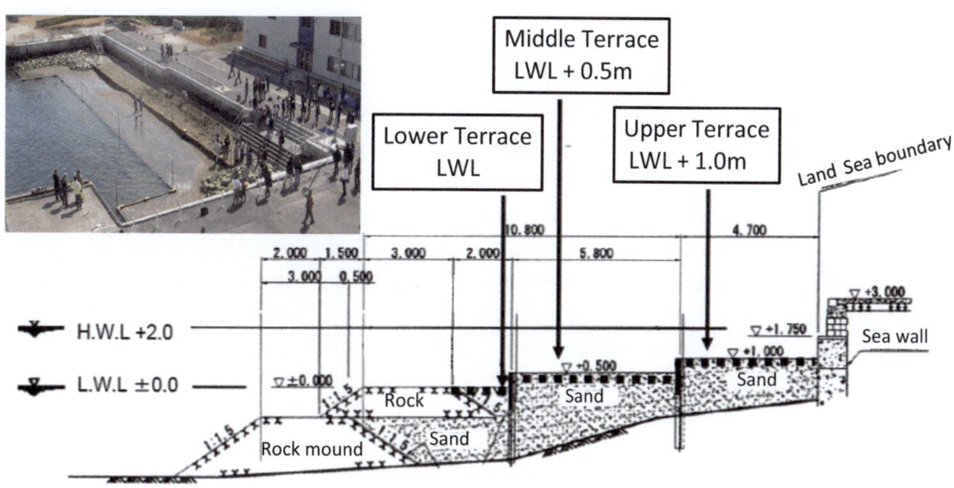

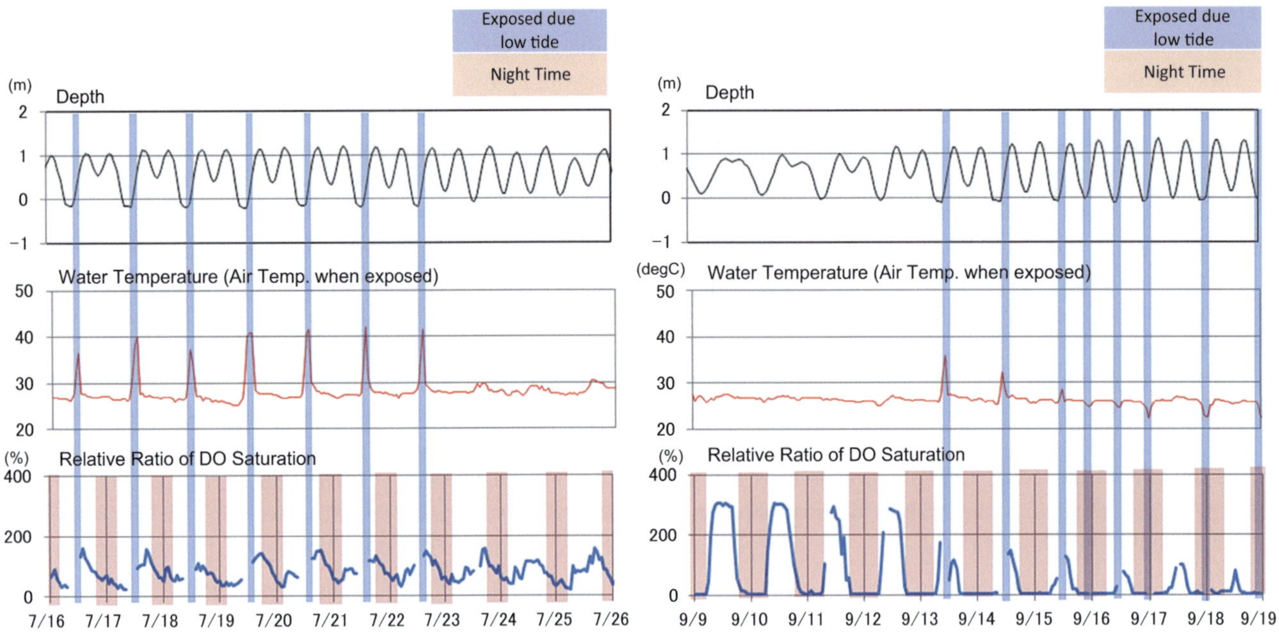

Fig. 2.40 Detailed observations of DO concentration on the upper terrace of the flat during early summer (*left*) and late summer (*right*) (Morita et al. 2009)

Gobies (*Acanthogobius flavimanus*) nest in soft saline seafloor in 6–8-m-deep water in winter; hatched larvae swim up to the upstream limit of the estuarine environment and settle on sandy shallow seafloor in spring. The fish grow to 6–8-cm long in early summer and up to 15–20 cm in autumn. These seasons are ideal for easy public fishing. Most of the senior residents, not only of Tokyo Bay but also almost the entire coast of Japan, experienced goby fishing in childhood. Hence, this activity can be considered to be part of the vanishing culture of coastal residents in Japan.

Some goby fishing events have been organised to remind people of the culture and attract interest for environmental restoration of the bay. In the 1950s, Hiyama et al. (1954) validated the reliability of the fishing censuses conducted by mixed participants, from beginners to experts. He found that the number of fish caught was significantly different, but there appeared to be no significant differences in the size fraction of the catch. Thus, we used the size of the caught fish as an indicator. This measure shows indications of changes in early growth and established habitat patterns. Some groups appear to have a stationary habitat in canals or seaside parks. Furthermore, the nesting places also shifted. While still more scientific research is required for the qualitative assessment of the change and its relation with environmental status, the preliminary study shows a large potential of goby habitat information as a useful indicator of coastal environmental change.

In 2012, the goby census was promoted by NILIM as a part of the *Edo-mae* project, which is an unofficial working group activity for the promotion of monitoring and restoration of goby habitats in Tokyo Bay (www.meic.go.jp/mahaze/). The goby census was conducted between July 1 and September 30. Participants in the census were requested to send monitoring sheets collecting total length data by facsimile to the census secretariat (Fig. 2.41).

In total, 239 data sheets were collected with 137 data points. The average fish length was 91 mm for July ($N = 456$ from 1,384 catch records), 101 mm for August ($N = 1{,}040$ from 1966 catch records) and 120 mm for September ($N = 1{,}727$ from 5,292 catch records). The average catch per person per hour for all census data was 20.1 individuals.

These data provide a standard for assessing the local datasets (Table 2.3, Fig. 2.42). For example, in Hirakata Bay (a: in Fig. 2.42) and the Tama River (c: in Fig. 2.42), relatively small gobies were captured. These areas are remaining natural shallows and tidal flats in a brackish-water environment. This is considered to be the habitat of juvenile gobies. Thus, these environments could be a networked nursery in the bay and should be well protected from environmental degradation of bay water. Conversely, at Yokohama Harbour (b: in Fig. 2.42), the Ara River (d: in Fig. 2.42) and Chiba Harbour (f: in Fig. 2.42), relatively rapid growth was observed and bigger gobies were captured. These areas are enclosed and face the

Fig. 2.41 Image of goby census incorporating public participation (*top left* and *right*, fishing events; *bottom left*, measurement of fished gobies; *bottom right*, data-gathering fax sheet)

Table 2.3 Notation used in Fig 2.42 and average size of gobies

Name of Place	Plot in Fig. 2.5	Average size (mm)		
		July	August	September
Hirakata Bay	a	77	85	116
Yokohama Harbour	b	NA	103	122
Tama River	c	76	NA	114
Ara River	d	90	105	131
Edo River	e	99	105	112
Chiba Harbour	f	NA	109	127
Banzu tidal flat	g	87	94	115

hypoxic zone in the bay. Thus, fish are likely to be trapped in the area without input from and connection to the bay environment. Consistent with this hypothesis, at the Tama River (c: in Fig. 2.42) and the Banzu tidal flat (g: in n), no significant growth of gobies was recorded. Both these areas are associated with the remaining tidal flats in the bay. These areas are a part of a good network with the tidal flats, and movement from upstream (migration of small gobies) and emigration to the front of the tidal flat (migration of large gobies) cover the actual growth rate of individual gobies.

These tentative results suggest the existence of several possible negative pressures on goby habitat: (1) lack of habitat for juveniles in canals, (2) fishing pressure in canals, (3) low DO in intermediate and deep areas during summer, (4) lack of habitat for nesting and (5) food shortage in intermediate areas. If it is possible to reverse these pressures, they could be good countermeasures for goby conservation programmes.

Good collaboration and the participation of the general public are a vital part of the implementation of environmental measures in the bay. The establishment of a system to

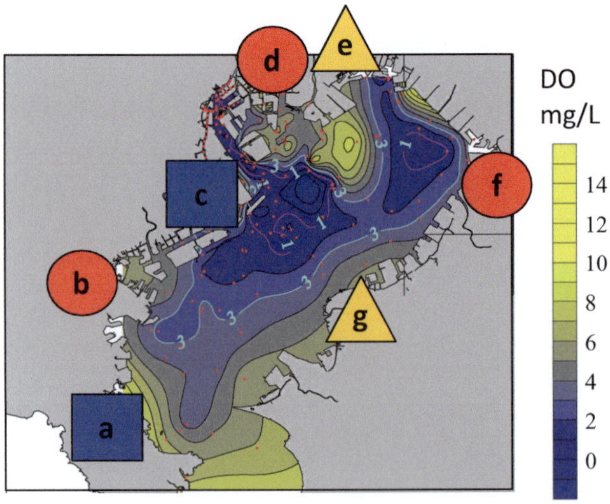

Fig. 2.42 A conceptual map of the relation between goby growth rate and environmental constitution (hypoxic environment) of Tokyo Bay

accommodate public participation and provide good incentives for participants is also an important future task.

2.6.5 Summary of Future Directions

Important and practical future issues are listed below:

Sustain a monitoring campaign in Tokyo Bay.

- Monitoring should be performed not only by specialists but public participation should also be promoted.
- Items to be monitored can be adaptively managed to obtain the best practical results.
- The output of the monitoring should be provided to the public in a transparent manner. The production of an environmental map with a workshop-type information exchange mechanism is an example of good practice.

Create and maintain habitat for benthos, sessile organisms and fish larvae in Tokyo Bay.

- Canals are frontline areas for habitat restoration. Even if each habitat is small, they can be networked along the canal.
- Restoring shallows in the bay is a crucial part of the action plan. A good understanding of the potential of the bay as habitat and an ecosystem-based approach to harmonise with the holistic system of the bay is important.

Establishment of public participation mechanisms.

- Enhance people's awareness and understanding of the environmental problems in Tokyo Bay.
- Let people act for what they wish; show/explain the results obtained by their actions.
- When people get involved, the bay environmental restoration becomes a part of their own consciousness.
- Monitoring is a good tool for promoting awareness and understanding; habitat creation is a good tool for achieving action.

In conclusion, the establishment of a new holistic approach for sustainable management of the bay (ICM: integrated coastal management) to facilitate monitoring, action and participation is urgently required for the environmental restoration of Tokyo Bay.

2.7 Concluding Remarks

2.7.1 Change and Status in Environment of Tokyo Bay

The population in the Tokyo Bay has increased quite rapidly, and shoreline changes due to land reclamation were significant during the last century. The population has almost quadrupled, and a land area of 300 km^2 has been reclaimed since the early twentieth century. Fisheries have also altered drastically, and the *Edo-mae* culture is diminishing.

Changes in the circulation pattern were triggered by changes in the freshwater discharge due to the civilisation of the Tokyo Bay watershed. Eutrophication of the bay peaked during the 1980s, and thereafter, the total nutrient load has become almost half of that during the worst stage. Nonetheless, indicators of eutrophication (red tide, hypoxia/blue tide and fish mortality) are observed.

2.7.2 Responses and Summary of Future Directions

Load control and habitat restoration are ongoing under the initiative of the 'Tokyo Bay Renaissance Project'. In particular, the implementation of sewage treatment systems has significantly contributed to minimising nutrient input; however, TP influence from the bottom sediment continues. In addition to eutrophication, the shallow-water region has rich biodiversity of benthos and sessile organisms. Thus, ecosystem-based management to link habitats and surrounding environments is a crucial concern in the restoration of the Tokyo Bay environment.

Important practical future directions are (1) to sustain the monitoring campaign with the cooperation of various stake holders; (2) to create and maintain habitat for benthos, sessile organisms and fish larvae using an ecosystem approach; and (3) to establish mechanisms for public participation (Boxes 2.1 and 2.2).

Box 2.1 Memory of Tokyo Bay as a rich habitat of Goby

Tokyo Bay was a rich habitat of goby (*Acanthogobius flavimanus*). It has nest in soft saline bed in 6–8-m-deep water in winter, and hatched larvae swim up to the upstream limit of estuarine environment and settle on sandy shallow bed in spring. It becomes 6–8 cm in early summer and growing up to 15–20 cm in autumn. These seasons are ideal for easy fishing for public. Most of the senior residents, not only in Tokyo Bay but also almost all around Japan, had experience of goby fishing in childhood. So, it can be said as one of the vanishing cultures for coastal residents in Japan. Based on the total number of fishermen and average catch, in the 1960s, total catch of the bay estimated as 100 million individuals a year. The number of catch decreased by factor 1/10 by 20 years. In 2000s, total catch can be one million. It is thought as a critical number to sustain its family. Now, monitoring campaigns are done to draw public awareness on this issue.

Fig. Box 2.1 Goby fishing boats at Tokyo Bay in 1960s. Boats are lined up on the targeted depth (6–8 m) where gobies nest exists (photo courtesy of Y Sawada)

Box 2.2 Historical propagation of reclamation in Tokyo Bay

Reclamation in Tokyo Bay was already being done in the Edo Period (1603–1867). The area of reclaimed land began increasing in the decade of 1955–1964 and peaked during the 1965–1974 years. Reclamation along the Keihin Canal also began in the Edo Period. In the early days of the Meiji era (1868–1912), the areas along the present Tokaido Line were the coastlines. In the Taisho era (1912–1926), the reclamation of the areas north of Shinagawa Station began, and Shibaura and its environs already had the topography seen today. On the other hand, in areas south of Shinagawa Station, the coastline was about where today's Keihin Kyuko Line is situated, where reclamation began in earnest in 1955 and after. The 'Third Revised Tokyo Port Plan' drawn up in 1976 defined its main purposes to be the conservation and recovery of nature and the multipurpose use of reclaimed land as a solution to urban problems. The present seaside parks, including Central Oi Wharf Seaside Park, Tokyo Port Wild Bird Park, Jonanjima Seaside Park and Omori Country Beach Park, have been built around the Keihin Canal by the Tokyo Metropolitan Government and Tokyo's wards to provide citizens with places to enjoy on beach outings.

(continued)

Box 2.2 (continued)

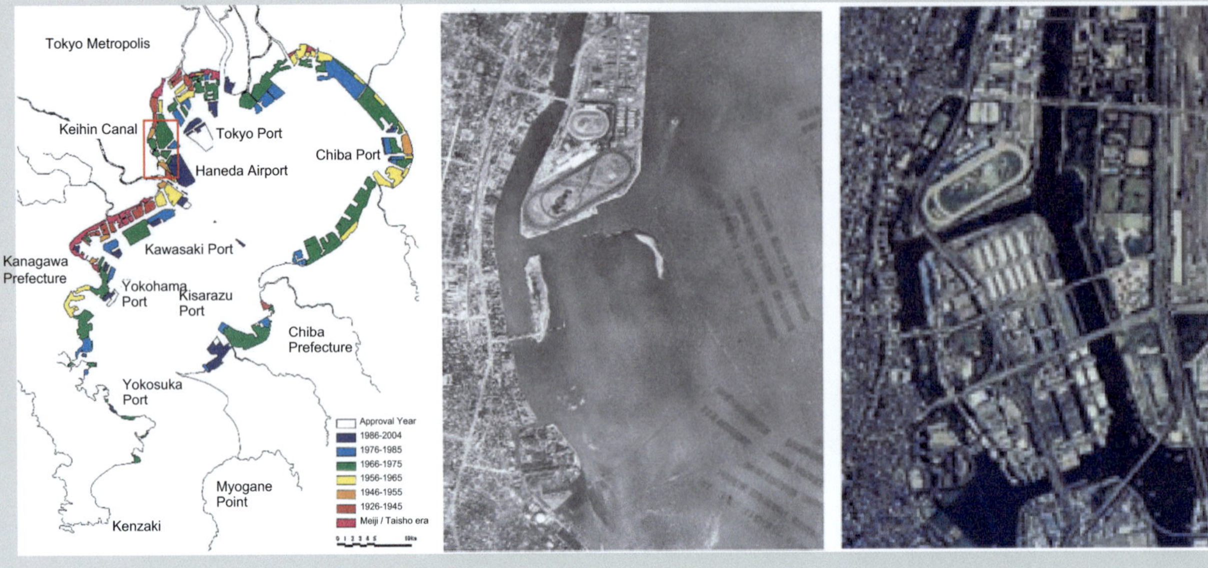

Fig. Box 2.2 History of reclamation in Tokyo Bay (left figure: reclamation approval after 'White Paper on the Tokyo Metropolis, 2005', Ministry of Land, Infrastructure, Transport and Tourism. Mid and right photos: Oi Wharf and its environs in 1956 (*left*) and 1997 by aerial photos taken by the Geographical Survey Institute)

References

Ando H, Kashiwagi N, Ninomiya K, Ogura H, Kawai T (2005) Changes in the state of water pollution in Tokyo bay since 1980. Annual report of the Tokyo Metropolitan Research Institute for Environmental Protection, Tokyo, pp 141–150

Asai T, Ozasa H, Murakami K (1997) Effects of physical parameter on sessile assemblage. Technical Note of Port and Harbour Research Institute, Japan, 880, 27 p

Asakura K (1907) Red tide in Tokyo Bay. J Meteorol Soc Jpn 34:451 p (in Japanese)

Central Council for Environment (2004) Direction of 6th water quality control by total quantity, Report of the experts committee for water quality control by total quantity, Japan, 13 p

Committee of editing board for historical record of Tokyo Inner Bay fisheries (1971) Historical Record of Tokyo Inner Bay Fisheries, Publishing Party of Historical Record of Tokyo Inner Bay Fisheries, 853 p (In Japanese)

Furota T (2005) Importance of network system among benthic-animal local populations in bay waters. Bulletin of Fisheries Research Agency, Annex 3, Japan, pp 35–46

Furota T (1980) Seasonal variation of phytoplankton standing stocks in temperate embayment. Bull Plankton Soc Jpn 27:63–73 (in Japanese)

Furota T (2005) Importance of network system among benthic-animal local populations in bay waters, Bulletin of Fisheries Reserach Agency, Annex 3, Japan, 35-46.

Furukawa K (2011) Satoumi at work: an urban project in Tokyo, In: United Nations University Institute of Advanced Studies Operating Unit Ishikawa/Kanazawa (2011). Biological and Cultural Diversity in Coastal Communities, Secretariat of the Convention on Biological Diversity, Montyreal, Technical Series 61, pp 54–61

Furukawa K, Ishimaru S (2012) Experiment on transparency monitoring in monitoring campaign of Tokyo Bay. In: Proceedings of JACZ, Japan (9-1-1)-(9-1-4) (in Japanese)

Furukawa K, Okada T (2006) Tokyo bay: its environmental status–past, present, and future". In: Wolanski E (ed) The environment in Asia Pacific harbours. Springer, Dordrecht, pp 35–45

Hasegawa K, Hayashi T (2009) Nutrient status and *nori* (*Porphyra*) aquaculture in Tokyo Bay. Aquabiology 31(2):161–164

Hinata H (2006) Effects of oceanic water intrusion on the Tokyo Bay environment. In: Wolanski E (ed) The environment in Asia Pacific harbours. Springer, Dordrecht, pp 67–78

Hinata H, Furukawa K (2006) Ecological network linked by the planktonic larvae of the clam Ruditapes philippinarum in Tokyo Bay. In: Wolanski E (ed) The environment in Asia Pacific harbours. Springer, Dordrecht, pp 35–45

Hiyama Y, Kasuka T, Nose Y (1954) Size selection in fishing caused by size of hook and skill of Angler. Jpn J Ichthyology 2(3):134–137

Igarashi M, Furukawa K (2007) Characteristics of spatial distribution of benthos and sessile organisms at Tokyo Bay shore. Annu J Civil Eng Ocean 23:459–464

Kaizuka S (1993) Geographical and geological features of Tokyo Bay (in Japanese), Tsukiji Shokan, 211 p (in Japanese)

Kakino J (1986) Effect on shellfish of blue tide in Tokyo Bay. Aquacult Fish Port Eng 22:41–47 (in Japanese)

Kamimura S, Yoshida J, Okada T, Furukawa K (2011) Nationwide study on the sessile assemblage inhabiting coastal structures, Japan. Research report of National Institute for Land and Infrastructure Management, Japan, 44, 51 p

Ministry of Environment (2010) Mid-long term vision on enclosed sea. Report of round table conference on making a mid-long term vision on enclosed sea, Japan, 86 p

Morita K, Watanabe M, Furukawa K, Imamura H, Kameyama Y, Morohoshi K (2009) Habitat rehabilitation on environmental symbiosis type of seawall and its maintenance experiences under public-private partnership. Proc Civil Eng Ocean 25:987–992 (in Japanese with English abstract)

Nomura H (1998) Changes in red tide events and phytoplankton community composition in Tokyo Bay from the historical plankton

records in a period between 1907 and 1997. Oceanogr Jpn 7:159–178 (in Japanese)
Ogura N (1993) Tokyo Bay–its environmental changes (in Japanese). Koseisha Koseikaku, Tokyo, 193 pp (in Japanese)
Okamura K (1907) Red tide in Kisarazu coastal area. Jpn J Fish 2:1–5 (in Japanese)
Okubo A (1973) Effect of shoreline irregularities on streamwise dispersion in estuaries and other embayments. Neth J Sea Res 6:213–224
Oomura T, Morohoshi K, Hosokawa Y (2009) Benthic habitat restoration along seawall under co-benefit strategy with disaster prevention. Paper presented during the East Asian Seas Congress 2009. Manila, Philippines
Perillo GME (2009) Tidal courses: classification, origin and functionality. In: Perillo GME, Wolanski E, Cahoon DR, Brinson MM (eds) Coastal wetlands. Elsevier, Amsterdam, pp 185–209
PIANC (2003) Ecological and engineering guidelines for wetlands restoration in relation to the development, operation and maintenance of navigation infrastructures. EnviCom report of WG7, The World Association for Waterborne Transport Infrastructure, Belgium, 64 p
Sakurai N, Boon-Keng L, Kobayashi K, Inoue H, Furukawa K, Hayakawa O (2008) Applies an adaptive management in practices of rebuilding creatures' living spaces, revival of *Edo-Mae*. In: Proceedings of the 20th ocean engineering symposium, Japan
Satoh C, Furukawa K, Okada T (2006) Macrobenthic fauna as restoration objectives and its constraints in Keihin Canal, Japan. Annu J Civil Eng Ocean 22:211–216
Suzuki T, Matsuyama M, Nagashima H (1997) A numerical experiment on the formation of circulation and upwelling by Northeasterly wind and stratified Tokyo Bay. Bull Coast Oceanogr 35(1):99–108
Takao T, Okada T, Nakayama K, Furukawa K (2004) Seasonal variation of residence time of sea water in Tokyo Bay during 2002. Technical Note of National Institute for Land and Infrastructure Management, Montyreal, Japan 169, pp 1–78 (in Japanese)
Tokyo Bay Marine Environment Research Committee (2011) Tokyo Bay–restoration of relation between human and nature. Koseisha-Koseikaku, Tokyo, 399 p
Tokyo Bay Renaissance Promotion Council (2003) Action plan of the Tokyo Bay renaissance, Tokyo Bay Renaissance Promotion Council, Tokyo, 36 p (in Japanese)
Tokyo Bay Renaissance Promotion Council (2013) Action plan of the Tokyo Bay renaissance (2nd phase), Tokyo Bay Renaissance Promotion Council, Tokyo, 39 p (in Japanese)
Tokyo Metropolitan City (2011) Report of public water quality monitoring 2011, Tokyo metropolitan city, Tokyo (in Japanese)
Unoki S (1998) Rivers viewed from the sea. Oceanogr Jpn, Japan 5:327–332 (in Japanese)
Unoki S, Kishino M (1977) Currents and water exchange in Tokyo Bay. Technical report of Physical Oceanography Laboratory in Institute of Physical and Chemical Research 1, 1–86 (in Japanese)
Yanagi T (2006) A summary of "open-ocean originated phosphorus and nitrogen in the coastal sea". Bull Coast Oceanogr 43(2):101–103
Yanagi T (2007) *Sato-umi*, a new concept for coastal sea management. Terra Scientific Publishing Company, Tokyo, 86 p
Yanagi T, Onishi K (1999) Change of tide, tidal current, and sediment due to reclamation in Tokyo Bay. Oceanogr Jpn 8:411–415 (in Japanese)
Yoshikawa K, Motonaga Y (2006) Post-evaluative study of flood damage mitigation in the basin of gentle flowing rivers on a low-lying plain. J Jpn Soc Hydol Water Resour 19(4):267–278

Oligotrophication in the Seto Inland Sea

3

Tetsuo Yanagi

Abstract

Over the past 50 years, the Seto Inland Sea has experienced both eutrophication and oligotrophication. In terms of year-to-year variation, the relationship between the phosphorus load and the number of occurrences of red tides shows a clockwise hysteresis, while that between nutrient concentration and fish catch has a counterclockwise hysteresis. Such a difference is the result of the mutual interaction between the water quality and bottom sediment, that is, the organic matter accumulated in the bottom sediment during the period of eutrophication. Therefore, the number of red tides during the period of oligotrophication is greater than during eutrophication at the same phosphorus load due to the release of dissolved inorganic phosphorus from the bottom sediment. The fish catch during oligotrophication is less than during eutrophication at the same nutrient concentration due to the generation of hypoxia in the bottom layer during the stratification period.

Keywords

Eutrophication • Oligotrophication • Interaction between water quality and sediment • Hysteresis

3.1 Introduction

Since ancient times, people have enjoyed the Seto Inland Sea, one of the most beautiful seas not just in Japan but in the world. It is blessed with numerous islands, a mild climate, low precipitation, and abundant nature (Fig. 3.1). The Seto Inland Sea is the largest semi-enclosed coastal sea in Japan and is surrounded by three large islands, Honshu, Kyushu, and Shikoku, with more than 700 smaller islands within its area. The total shoreline is 6,868 km in length (Fig. 3.2). It is about 450 km from east to west and varies from 15 to 55 km north to south. It has an area of 23,203 km^2 with an average depth of 38 m. Accordingly, the volume of the Seto Inland Sea is approximately 881.5 km^3. The sea has many large and small straits, bays, and basins ("nadas" in Japanese). The Kii Channel (in the east) and the Bungo Channel (in the west) lie between the Seto Inland Sea and the Pacific Ocean, while the Kanmon Strait (in the west) lies between the Seto Inland Sea and the Japan Sea.

Except for Beppu Bay, Iyo-Nada, the Bungo Channel, Kii Channel, and Hibiki-Nada, almost all bays and nadas in the Seto Inland Sea are shallow and less than 40 m in depth (Fig. 3.3).

In this region, the annual average temperature and precipitation are 15 °C and 1,000–1,600 mm, respectively. This means that the region is relatively mild (low levels of precipitation and a narrow annual range of air temperature and humidity) in comparison with other regions in Japan. However, the mountainous areas around the Seto Inland Sea have greater precipitation with an annual rainfall of 2,000–3,000 mm. The amount of water discharged into the Seto Inland Sea from 669 river systems is as much as 50 billion m^3 a year.

T. Yanagi (✉)
International EMECS Center, Kobe, Japan
e-mail: tyanagi@riam.kyushu-u.ac.jp

T. Yanagi (ed.), *Eutrophication and Oligotrophication in Japanese Estuaries*, Estuaries of the World, DOI 10.1007/978-94-017-9915-7_3

Through all seasons, the wind direction in the coastal areas is normally against the shoreline as a land-sea breeze develops. On the other hand, the wind direction in land areas largely depends on their individual geographical features. The wind pattern in the Seto Inland Sea is characterized by its regularity.

The seasonal variation in water temperature and salinity in the Seto Inland Sea is greater than that of the Pacific Ocean because the condition of the water in the Seto Inland Sea is generally affected by weather conditions.

Fig. 3.1 Landscape view of the Seto Inland Sea (Photo provided by the Seto Inland Sea Conservation Association)

Water is exchanged between the Pacific Ocean and the Seto Inland Sea through the Kii Channel and the Bungo Channel, while between the Japan Sea and the Seto Inland Sea it is through the Kanmon Strait.

Maximum tidal range in the Seto Inland Sea is about 3 m in the spring tide and the tidal current in narrow straits, such as the Naruto and Hayasui Straits, is fast and complex and reaches 5–10 knots. Complex geographical features affect the unique characteristics of the tide and tidal current in the Seto Inland Sea.

3.2 Changes in the Environment

3.2.1 Population

The watershed area of the Seto Inland Sea spreads out over about 68,000 km^2, which corresponds to 18 % of the total land area of Japan. The population of the area is about 35 million or about 28 % of the total population of Japan, and its annual specific rate of increasing has been around 2.8 % which is similar to the average for Japan as a whole (see Fig. 3.4).

Further, as shown in Fig. 3.5, the population density is 643 persons/km^2, which is 1.6 times greater than the average for Japan at about 340 persons/km^2. This means that the population is highly concentrated in this area, especially in Osaka Prefecture, with Fukuoka Prefecture coming next.

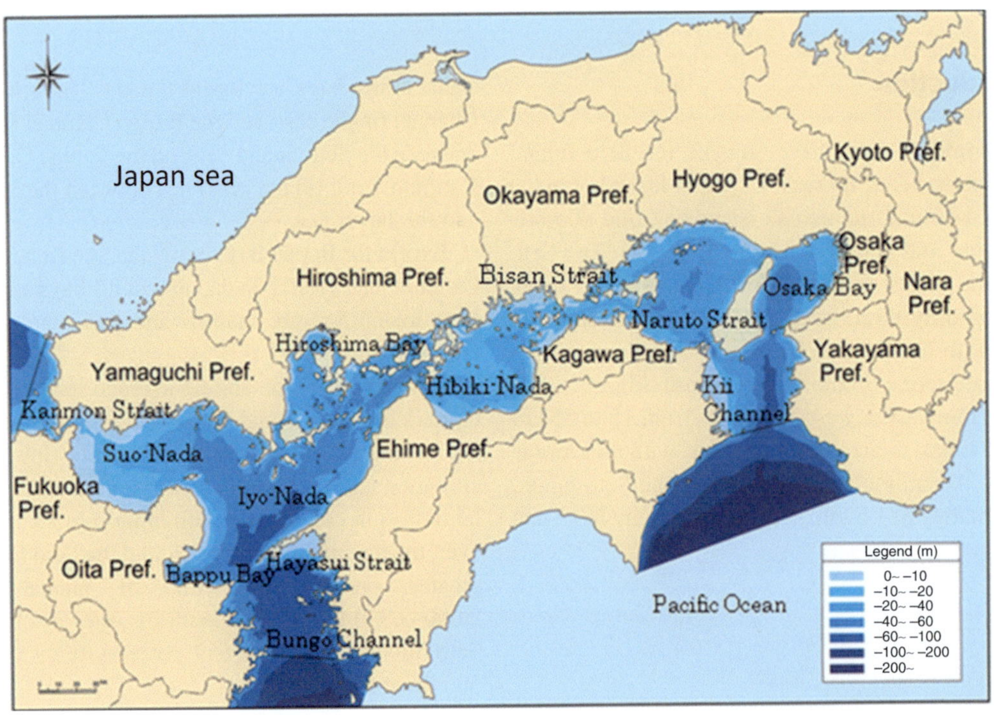

Fig. 3.2 Bays, nadas, straits, and channels in the Seto Inland Sea

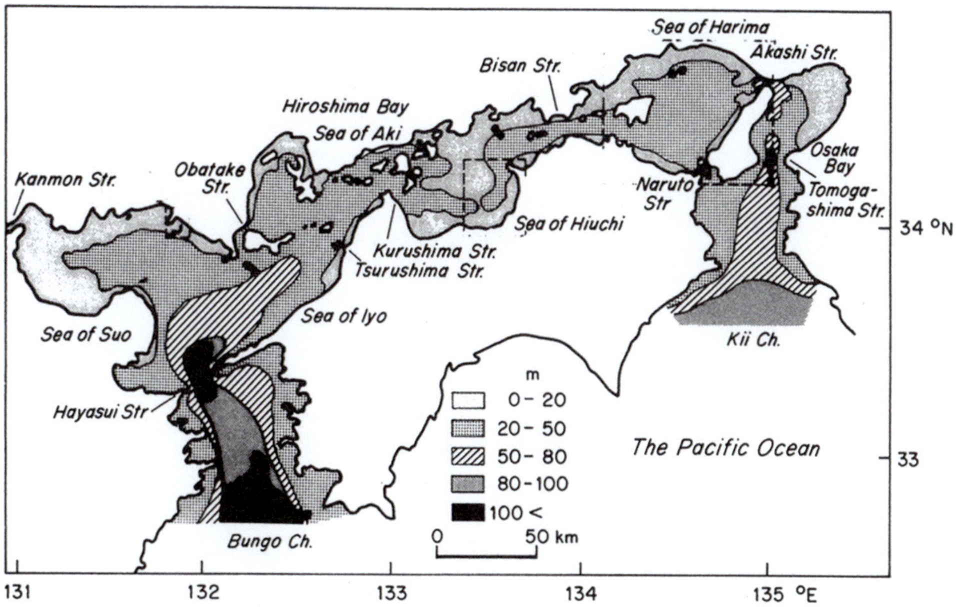

Fig. 3.3 Water depth in the Seto Inland Sea

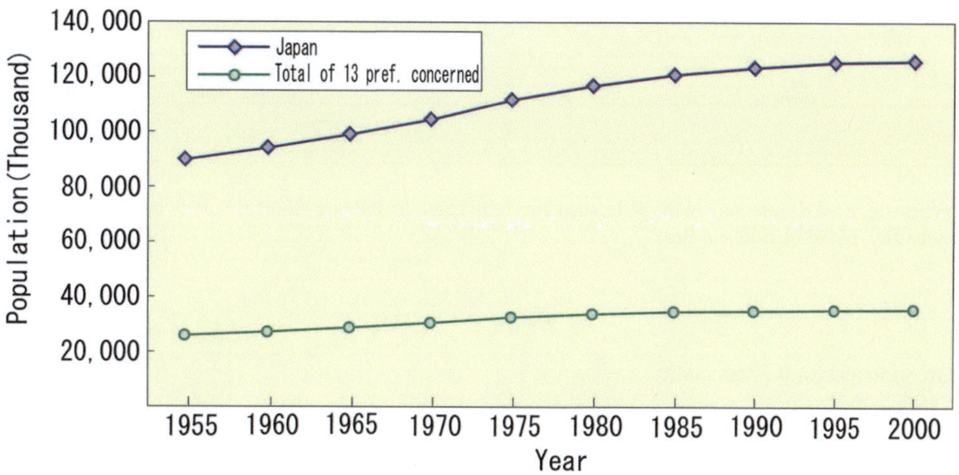

Fig. 3.4 Population of Japan as a whole and of the prefectures located in the watershed area of the Seto Inland Sea

3.2.2 Industry

The gross product of the 13 prefectures around the Seto Inland Sea was 129.8 trillion yen in 2002. The gross product from 1965 to 1995 had increased steadily, with an annual rate of increase of about 6.53 %. However, since 1996 the increase has not been very distinct and the level of production has become reasonably constant (Fig. 3.6).

With respect to the composition of industry in 1965, primary industries accounted for 7.4 % of the total, secondary industries represented 40.0 %, and tertiary industries 52.6 %. From 1965 to 1995, the ratio of secondary industries decreased by 7.3 %, while tertiary industries increased by 13.6 %, and this trend has not changed very much since then. The ratio in 2002, for example, was 0.85 % for primary, 25.3 % for secondary, and 73.8 % for tertiary industries.

The Seto Inland Sea offers advantageous conditions for industrial development, such as long shallow coasts suitable for reclaiming land for construction of factories and a population of some 30 million people living along the coastal areas. In the period of rapid economic growth of the 1960s and 1970s, heavy and chemical manufacturing industries grew up around the Seto Inland Sea, responding to the national government's "Income Doubling Plan". The construction of factories rapidly increased in the latter half of the 1960s after the New Industry City Law was enacted in 1963. At present, manufacturing industries in the area account for more than 30 % of the total production capacity for Japan.

Fig. 3.5 Population density of each prefecture in the watershed area of the Seto Inland Sea (Source: Survey conducted by the Association for the Environmental Conservation of the Seto Inland Sea)

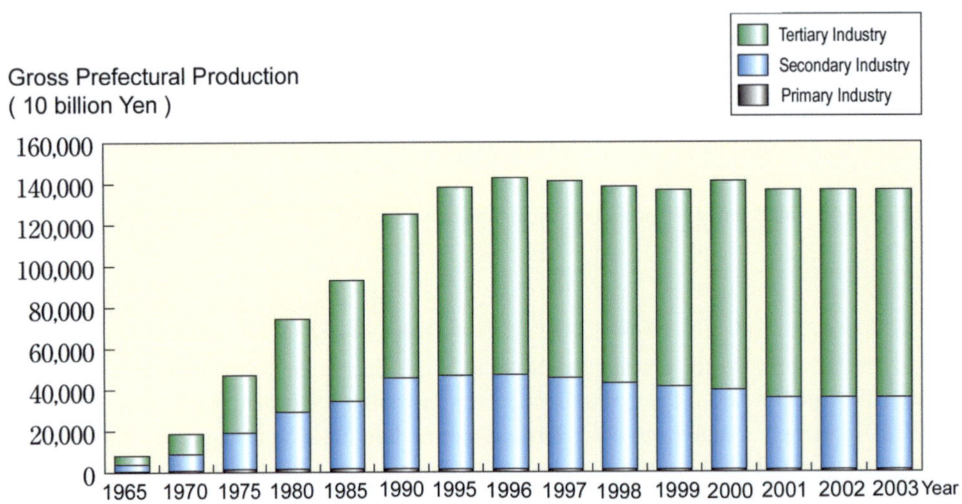

Fig. 3.6 Trends in and ratio of gross product in the 13 prefectures around the Seto Inland Sea (Source: Census for fiscal year of 2005)

There are basic manufacturing industries such as steel, oil refinery, and petrochemical industries in 13 prefectures around the Seto Inland Sea as shown in Fig. 3.7.

Trends in total shipping are shown in Fig. 3.8. The volume of shipping in this area has increased up to 1990, but since then it has changed and become rather stable. This trend is similar to that of Japan as a whole.

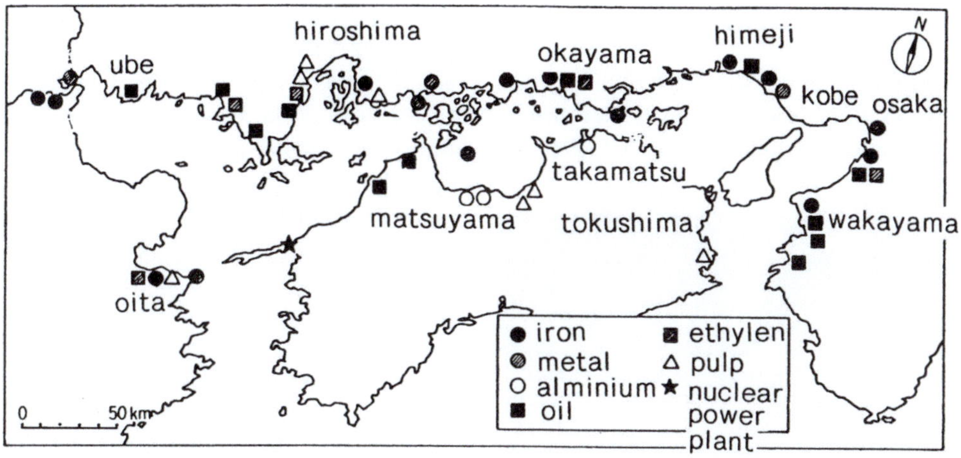

Fig. 3.7 Industrial areas and large cities around the Seto Inland Sea

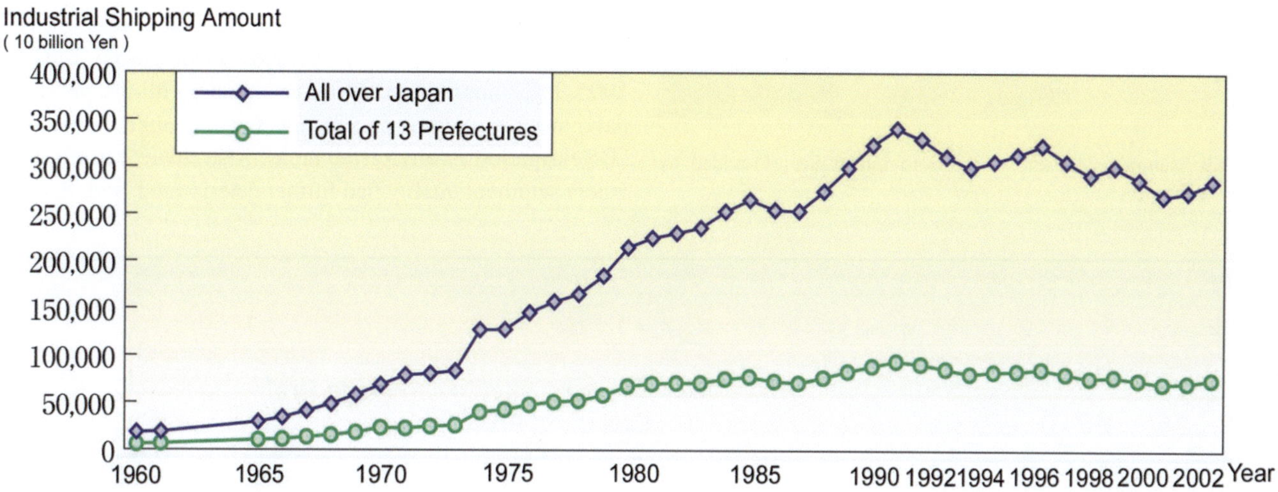

Fig. 3.8 Trends in manufacturing shipments in the prefectures concerned (Source: Ministry of International Trade and Industry)

3.2.3 Coastline

Since ancient times, the coastline of the Seto Inland Sea has been changed by reclamation for agriculture and salt farm land because of its shallow depth. The rapid industrialization that was undertaken from the 1960s required a great deal of reclamation, and, as a result, the natural coastline is only 37 % of the total coastline of the Seto Inland Sea at present (Fig. 3.9), which is less than the value for Japan as a whole (55.2 %).

Man-made beaches have been created to restore lost natural beaches. The state of the coastline along the Seto Inland Sea is shown in Fig. 3.10.

3.2.4 Dams

Many dams were constructed in the watershed area of the Seto Inland Sea and the number constructed was at its largest in 1966–1970, with about 10 per year as shown in Fig. 3.11. A total of about 600 dams are distributed in the watershed area of the Seto Inland Sea as shown in Fig. 3.12.

3.3 Status

3.3.1 Water Quality

The horizontal distributions of salinity, TP (total phosphorus), TN (total nitrogen), and Chl.*a* (chlorophyll *a*) in surface water during the summer of 2003 are shown in Figs. 3.13, 3.14, 3.15, and 3.16.

Low salinity was observed in the inner part of Osaka Bay and Hiroshima Bay, while high TP and TN were observed in the inner part of Osaka Bay, Hiroshima Bay, and Beppu Bay. High Chl.*a* was also observed in the inner part of Osaka Bay, the western part of the Kii Channel, the northern part of Harima-Nada, and Hiroshima Bay where high TP and TN were observed.

3.3.2 Sediment Quality

Generally, sediment quality in the sea is deteriorating, particularly in areas with stagnant water or a small tidal current. The horizontal distributions of mud content, TP, TN, and ORP (oxidation reduction potentials) in sediments from 2001 to 2005 are shown in Figs. 3.17, 3.18, 3.19, and 3.20. High TP and TN were observed in Osaka Bay, Harima-Nada, Hiuchi-Nada, Hiroshima Bay, the southwestern part of Suo-Nada, and Beppu Bay where mud content was high. Naturally, low ORP was observed in these areas.

The horizontal distributions of individuals, number of species, and the diversity index of macro-benthos in sediments are shown in Figs. 3.21, 3.22, and 3.23. High numbers of individuals and of macro-benthos species were observed at Bisan-Seto, in the western part of Bingo-Nada and Aki-Nada where TP and TN were low. Low biodiversity was observed in Osaka Bay, Harima-Nada, Hiuchi-Nada, Hiroshima Bay, the southwestern part of Suo-Nada, and Beppu Bay where TP and TN were high.

Observations of bottom sediment throughout the Seto Inland Sea have been conducted three times by the Ministry of Environment, Japan. The first one was from 1981 to 1987, the second from 1991 to 1996, and the third from 2001 to 2005. In comparing the most recent data with the previous data, no parameters except for ORP changed over the 20-year period in any of the areas. Also, there was no area where sediment quality had further deteriorated, and, in fact,

Fig. 3.9 Natural coast line in the Seto Inland Sea (Provided by Chugoku Press)

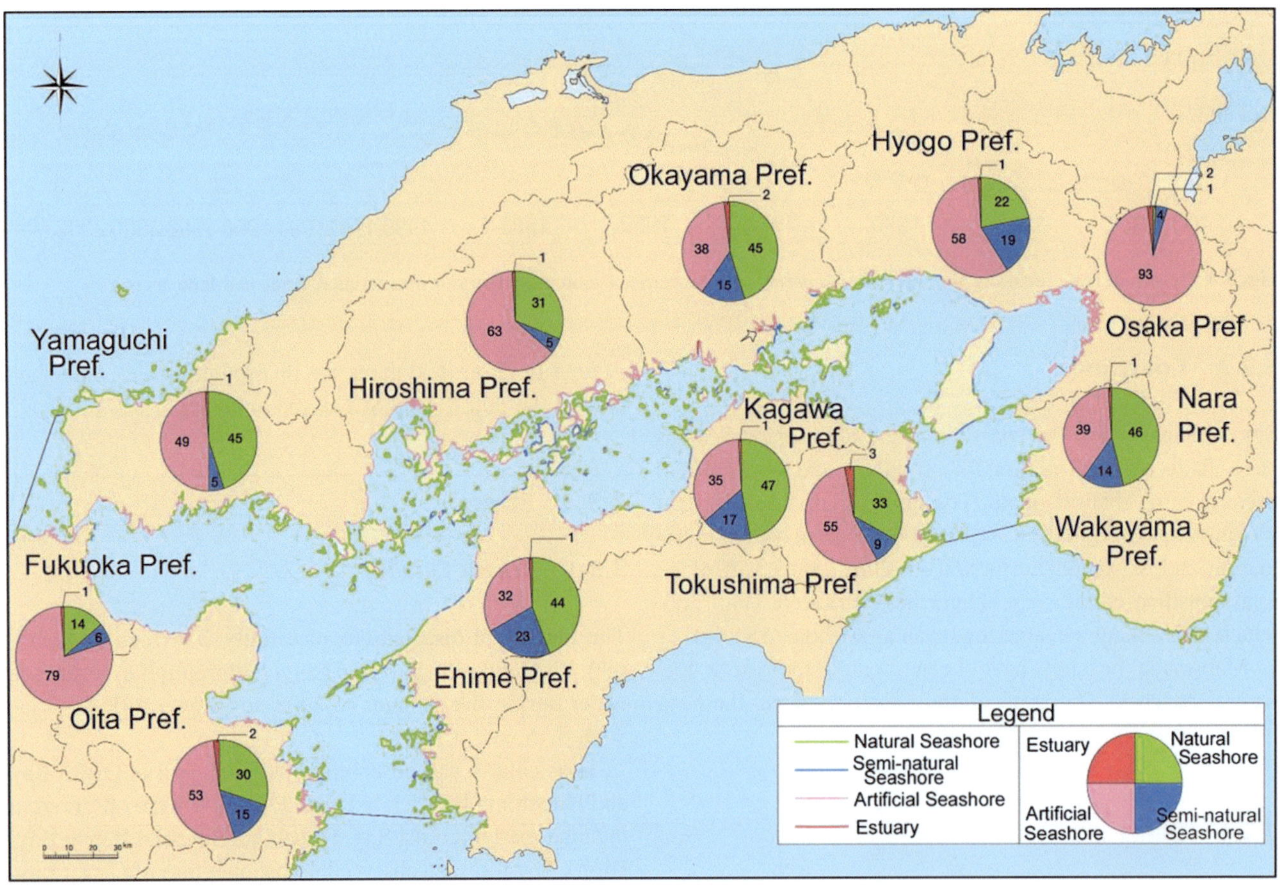

Fig. 3.10 State of the shore line along the Seto Inland Sea (Source: Environment Agency of Japan (fourth survey 1993); Regions of the Seto Inland Sea are based on the Law Concerning Special Measures. For Conservation of The Environment of The Seto Inland Sea)

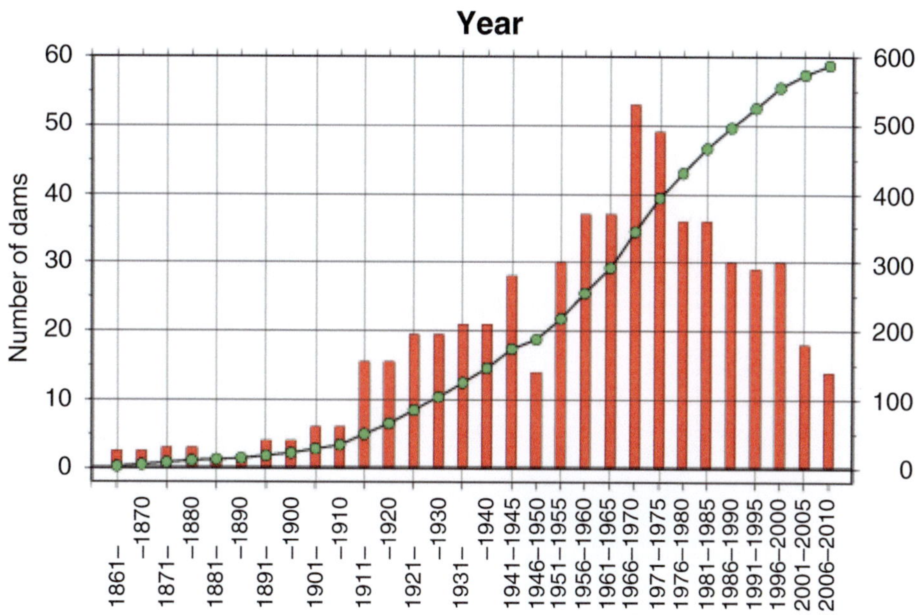

Fig. 3.11 Year-to-year variation in the construction and accumulation number of dams in the watershed area of the Seto Inland Sea

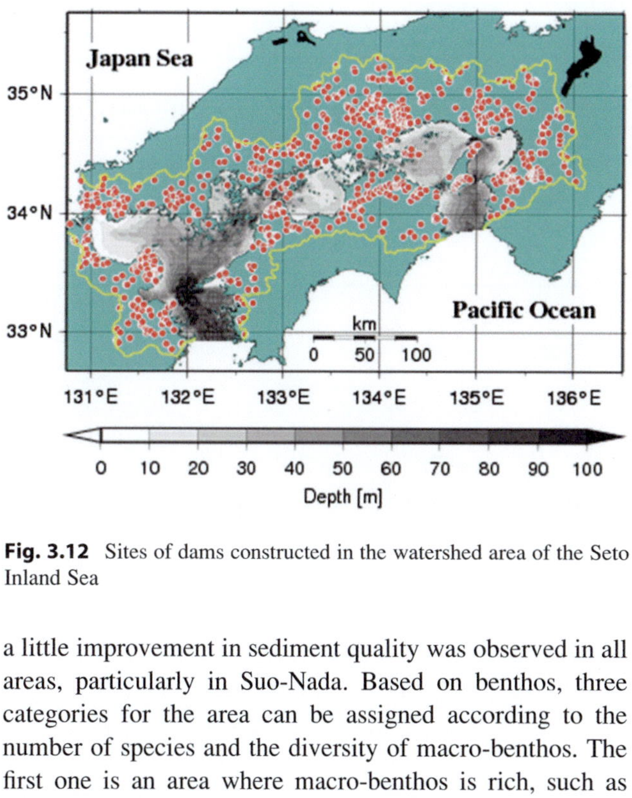

Fig. 3.12 Sites of dams constructed in the watershed area of the Seto Inland Sea

a little improvement in sediment quality was observed in all areas, particularly in Suo-Nada. Based on benthos, three categories for the area can be assigned according to the number of species and the diversity of macro-benthos. The first one is an area where macro-benthos is rich, such as Aki-Nada. The second is an area where macro-benthos is poor, such as in Beppu Bay, Hiroshima Bay, and Osaka Bay. The third is an area where the situation for macro-benthos is between the first two. In comparing the most recent data with the data previously obtained 10 years ago, statistically significant differences in individuals and the number of species of macro-benthos can be observed at Harima-Nada (decrease in individuals); Hiuchi-Nada (decrease in individuals and number of species); Kii Channel (decrease in number of species); Bisan-Seto (increase in number of species); Bungo Channel (increase in number of species); and Aki-Nada (increase in individuals and number of species). In comparing the results of the observation performed by the Fisheries Agency of Japan 20 years ago, it was thought that the number of species and the diversity of macro-benthos had decreased or shown no change in most areas. The main reason of such change is thought to be the effect of continuing hypoxia, as will be shown later.

Based on the correlation between water and sediment environments, it is thought that because of human activities and the small tidal current, it is easy to accumulate pollutants in the sea regions with a high content of sludge in the sediment, although the correlations among the quality of water and sediment and the environment of aquatic animals and plants are not completely understood. With respect to water quality in such sea regions, transparency is generally small during the summer.

Another important factor which controls of sediment quality is the effect of sea-sand mining lasting over a period of more than 50 years that stopped in 2006 in the Seto Inland Sea. Sea-sand mining was carried out over the whole area of the Seto Inland Sea where there was a sandy sea bottom, and this seriously affected the bottom environment due to changes in topography and the seagrass bed due to increases in turbidity. It especially affected the ecology related to sand eels (*Ammodytes personatus*) because the sandy bottom is

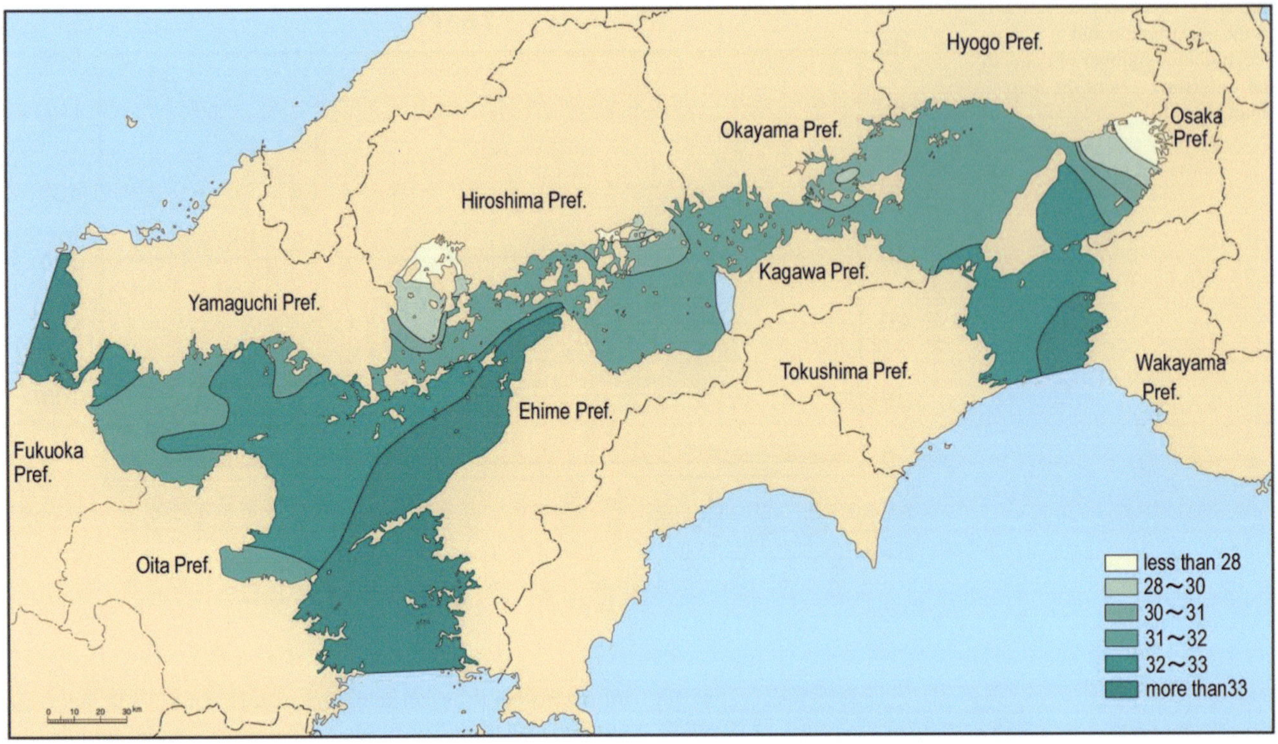

Fig. 3.13 Horizontal distribution of salinity in the surface layer during the summer season, 2003 (Source: Ministry of Environment of Japan 2004)

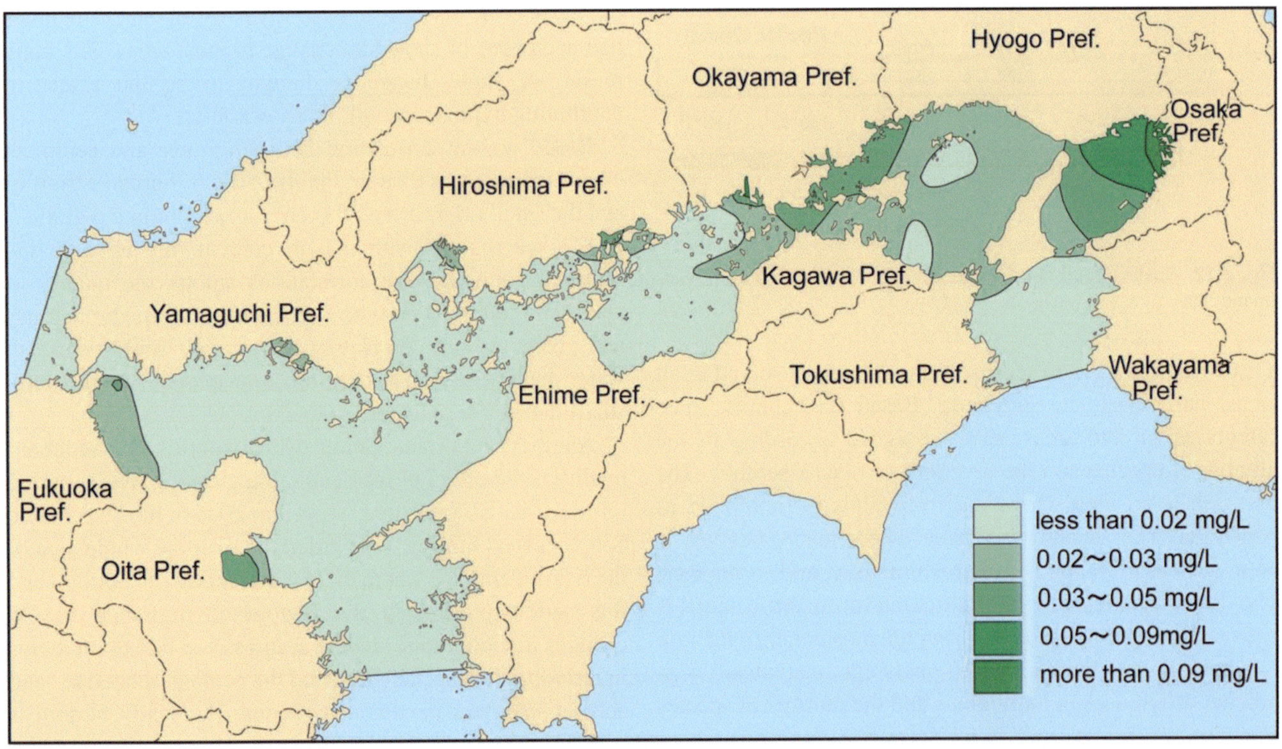

Fig. 3.14 Horizontal distribution of TP in the surface layer during the summer season, 2003 (Source: Ministry of Environment of Japan 2004)

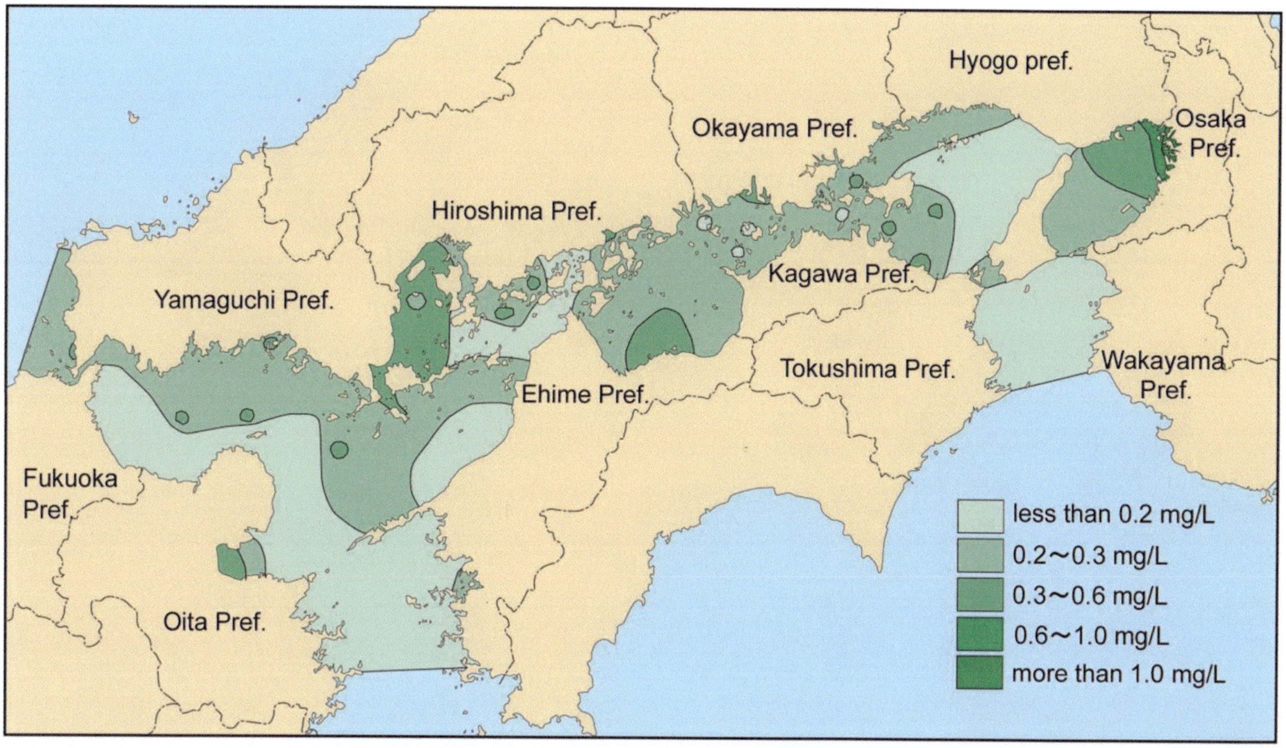

Fig. 3.15 Horizontal distribution of TN in the surface layer during the summer season, 2003 (Source: Ministry of Environment of Japan 2004)

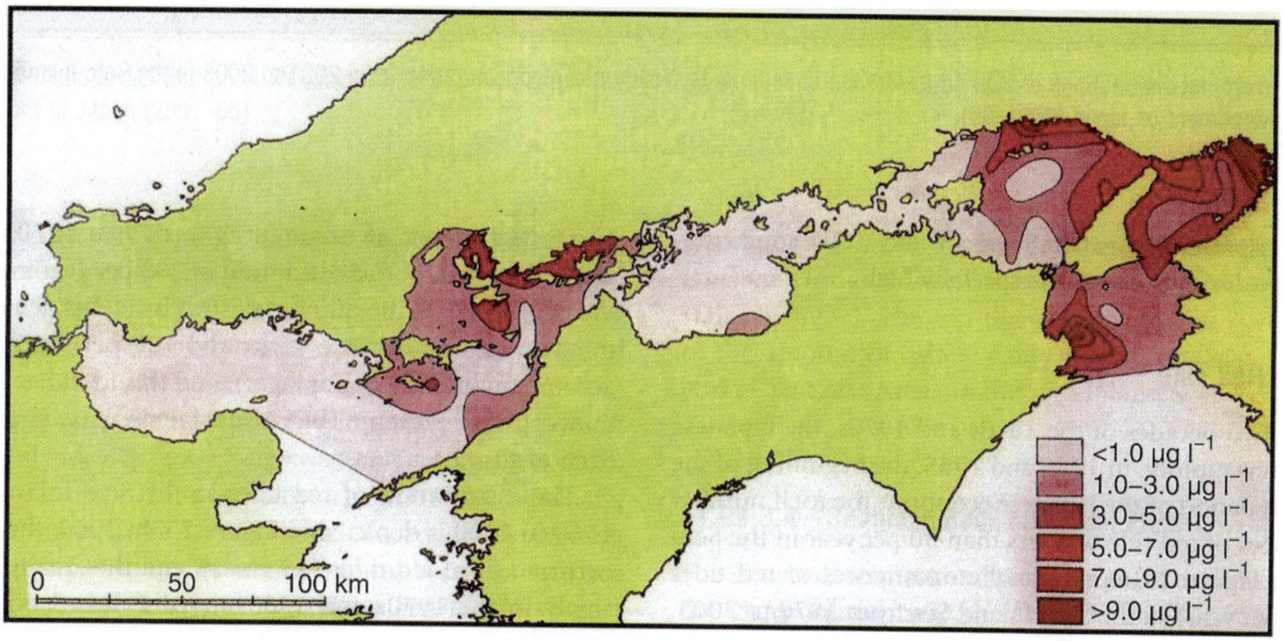

Fig. 3.16 Horizontal distribution of Chl.*a* in the surface layer during the summer season, 2003 (Source: Ministry of Land, Infrastructure and Transport Japan 2004)

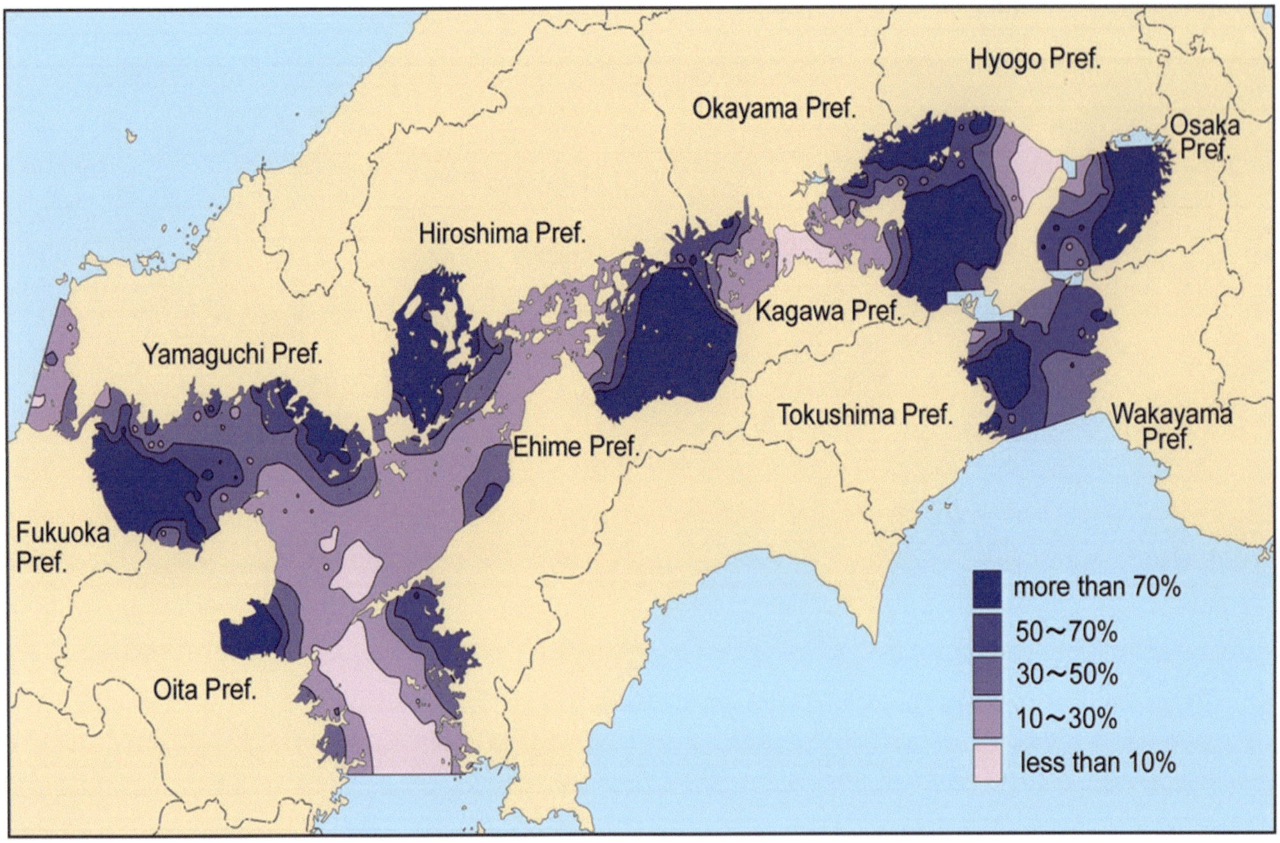

Fig. 3.17 Horizontal distribution of mud content (Source:Yearly mean concentration from 2001 to 2005 in the Seto Inland Sea; Ministry of Environment of Japan (2001–2005))

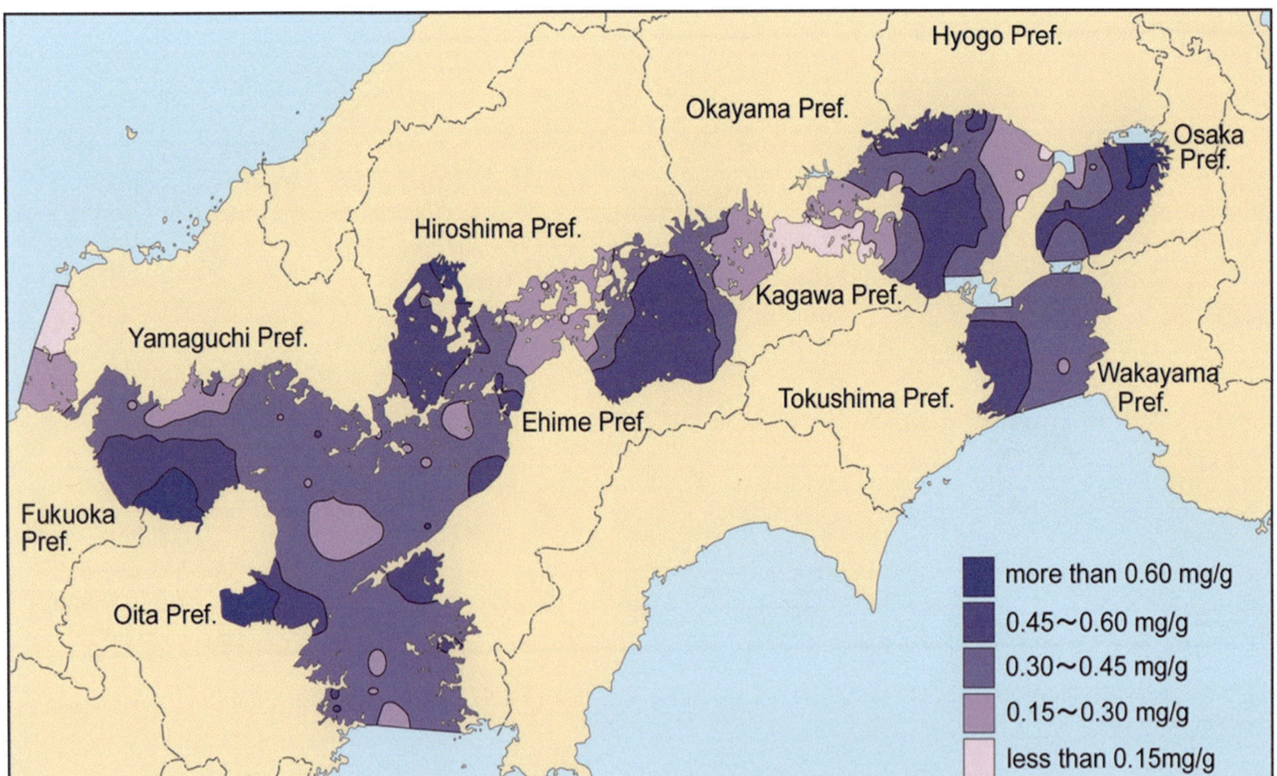

Fig. 3.18 Horizontal distribution of TP concentration in sediments (Source: Yearly mean concentration from 2001 to 2005 in the Seto Inland Sea; Ministry of Environment of Japan (2001–2005))

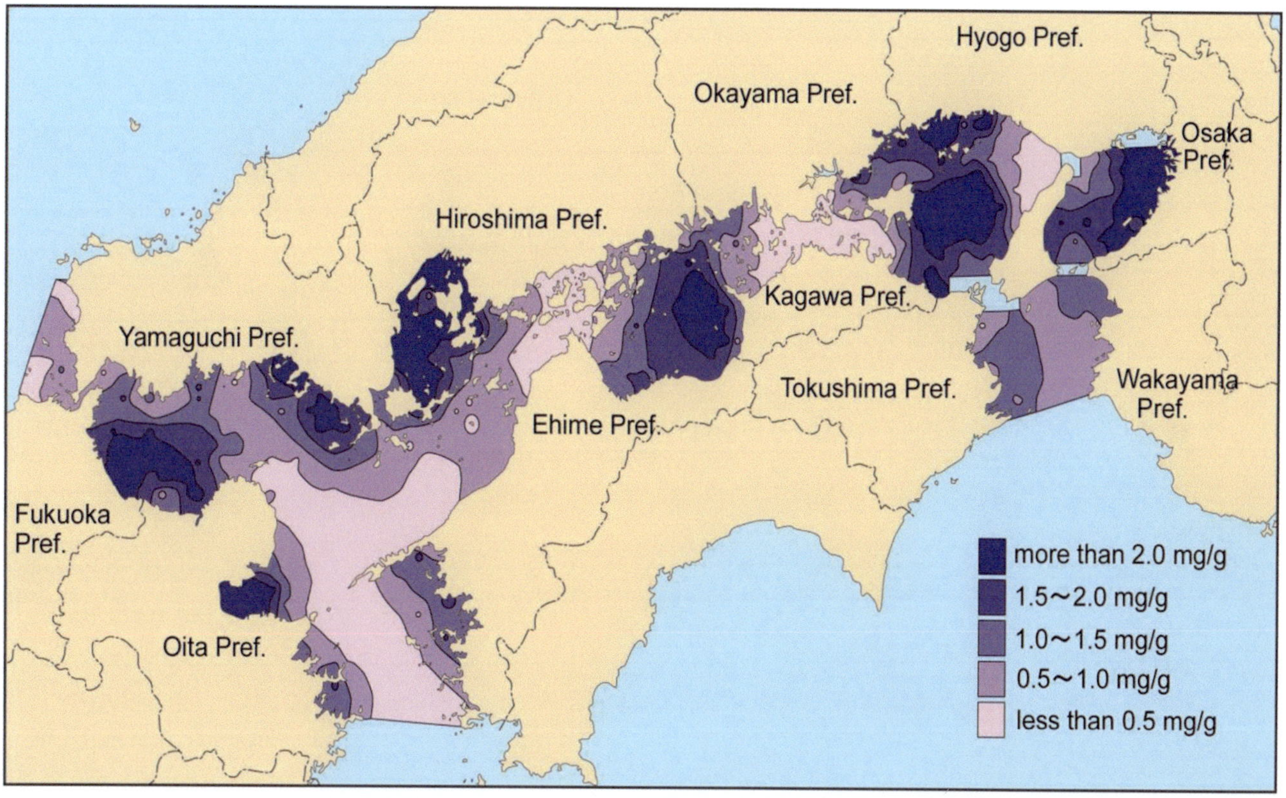

Fig. 3.19 Horizontal distribution of TN concentration in sediments (Source: Yearly mean concentration from 2001 to 2005 in the Seto Inland Sea; Ministry of Environment of Japan (2001–2005))

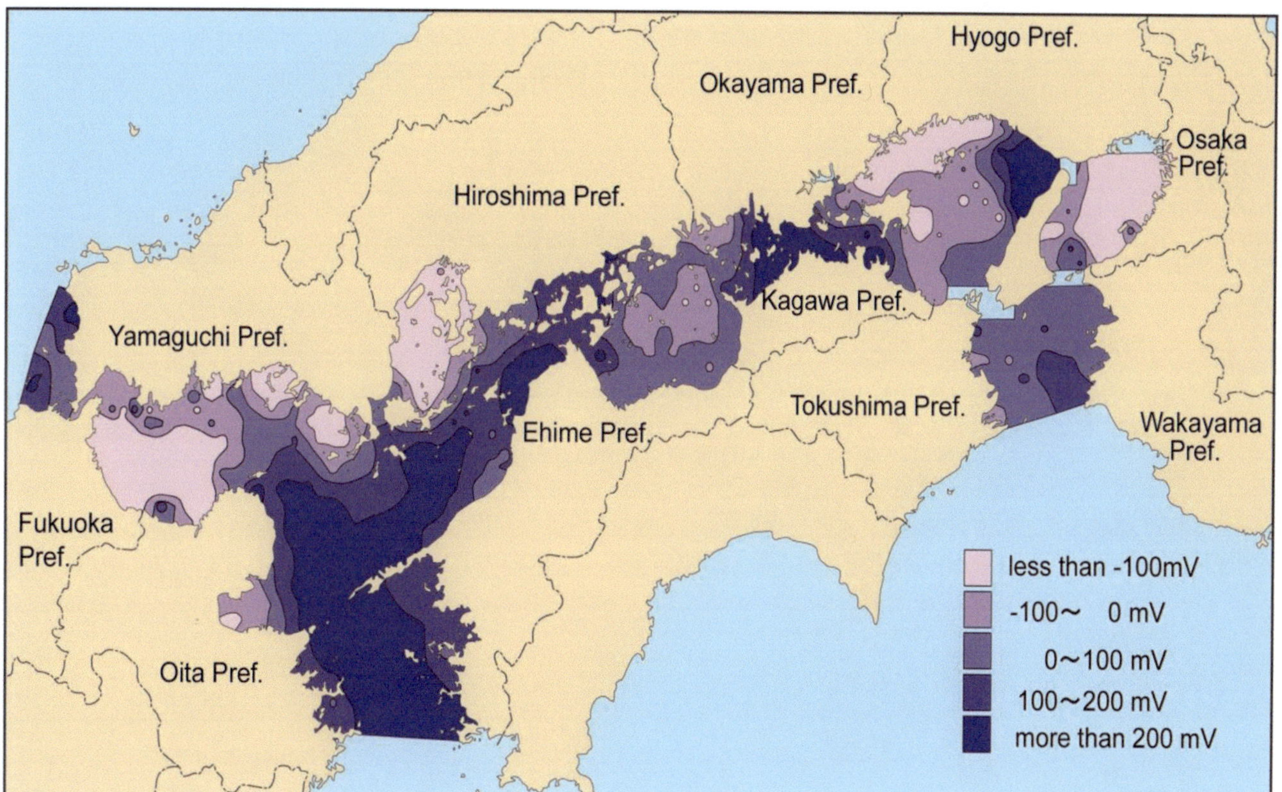

Fig. 3.20 Horizontal distribution of oxidation reduction potential in sediments (Source: Yearly mean concentration from 2001 to 2005 in the Seto Inland Sea; Ministry of Environment of Japan (2001–2005))

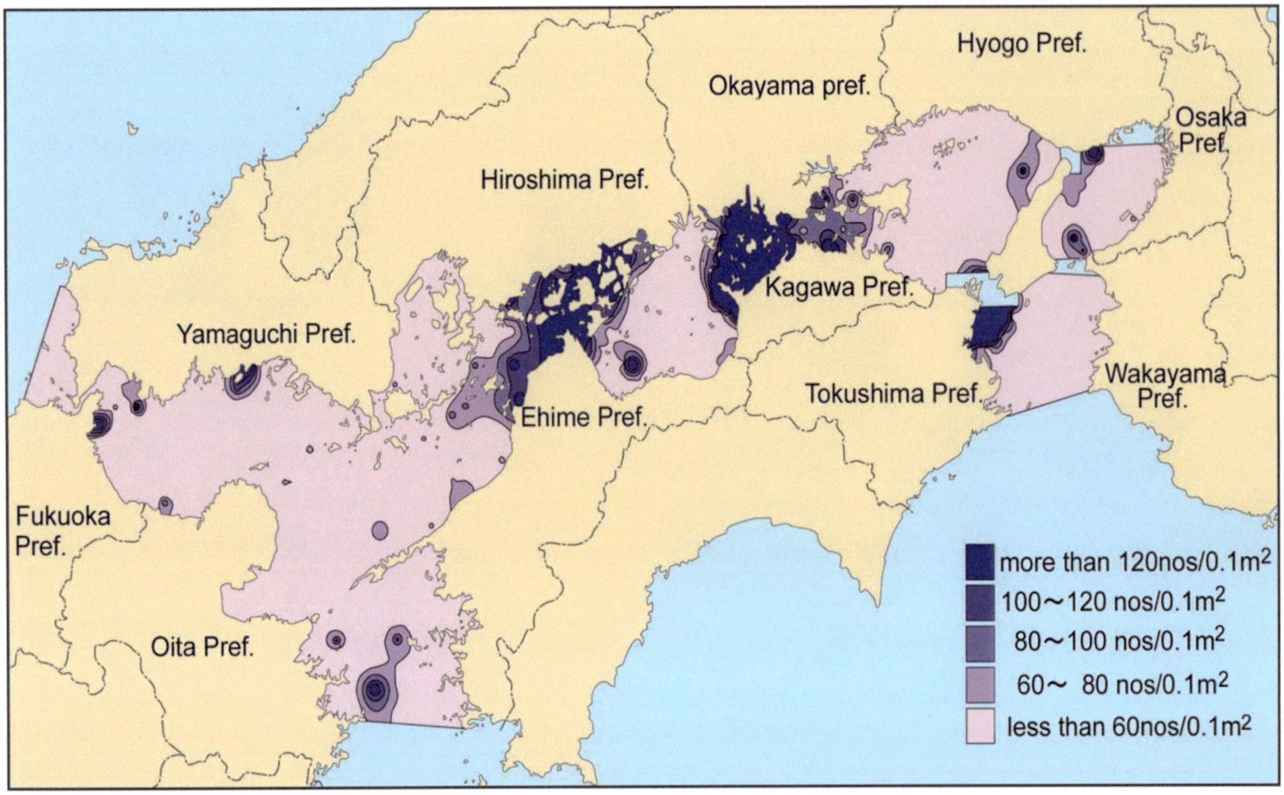

Fig. 3.21 Horizontal distribution of individual macro-benthos in sediments (nos = ind.) (Source: Yearly mean concentration from 2001 to 2005 in the Seto Inland Sea; Ministry of Environment of Japan (2001–2005))

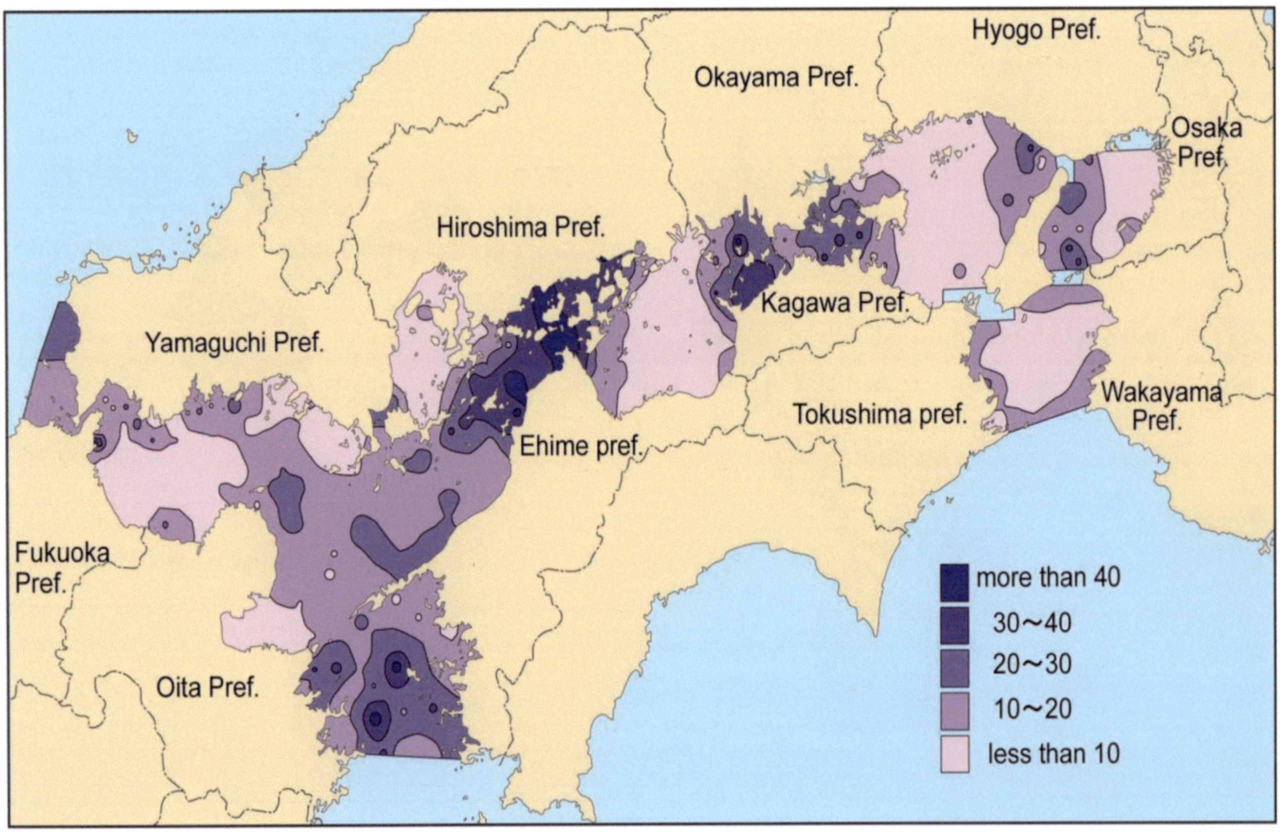

Fig. 3.22 Horizontal distribution of the number of species of macro-benthos in sediments (Source: Yearly mean concentration from 2001 to 2005 in the Seto Inland Sea; Ministry of Environment of Japan (2001–2005))

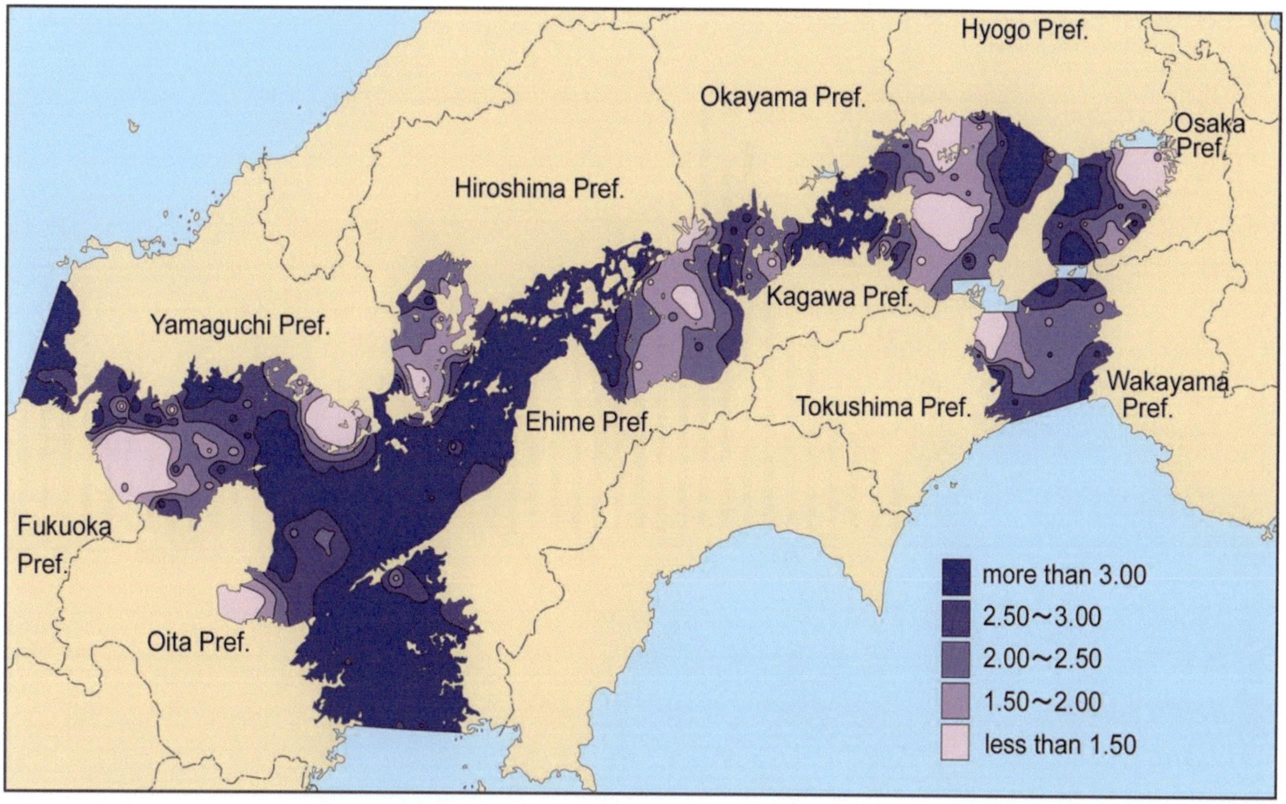

Fig. 3.23 Horizontal distribution of the Shannon-Weaver diversity index of macro-benthos in sediments (Source: Yearly mean concentration from 2001 to 2005 in the Seto Inland Sea; Ministry of Environment of Japan (2001–2005))

where sand eels sleep during the summer months. The effects of sea-sand mining are discussed in Yanagi (2008).

3.3.3 Red Tide

During the two decades of the 1960s and 1970s, the Japanese economy grew rapidly. From 1960 to 1965, the beginning of the period of high growth in the Japanese economy, the total number of occurrences of red tide was less than 50 per year in the Seto Inland Sea (Okaichi 2004). Figure 3.24 represents the number of red tides (incidents per year) from 1967 to 2008. The total number of occurrences was 50 in 1967, showing a marked increment to a maximum of 299 per year in 1976. After the peak in 1976, the number of such incidents demonstrated a trend to clearly decrease to around 100 per year (around 10 accompanying damage of fisheries), and this level has continued to the present time. The reason for this decrease will be discussed in Sect. 3.2.4.

Figure 3.24 is mainly based on the results as seen by local fishermen. Ishii et al. (2014) calculated a Red Tide Index (RTI; 10,000 ha·day) by multiplying the area and the period of red tide occurrences; their results are shown in Fig. 3.25. RTI in the narrow coastal area within 2 km of the coast has a similar magnitude of RTI to the wide offshore area, which suggests that red tides in the coastal area continue for a longer time than in the offshore area. The long-term trend in RTI is similar to that of the number of occurrences shown in Fig. 3.24.

The distribution of red tides in the Seto Inland Sea from 1960 to 2000 is depicted in Fig. 3.26. In the 1960s, there were few occurrences of red tides (18 cases), and the area involved was small. In the 1970s and 1980s, large-scale red tides occurred frequently, especially during the summer. In extreme cases, some red tides covered almost the whole area of the nadas and bays, such as Osaka Bay, Harima-Nada, Hiuchi-Nada, and Suo-Nada. By the 1990s and thereafter, the scale and period of red tides appears to have become smaller and shorter.

Typical microalgae causing noxious red tides in the Seto Inland Sea are *Chattonella antiqua* Hada (Ono), *Heterosigma akashiwo* (Hada) Hada ex Hara et Chihara (Raphidophyceae), *Noctiluca scintillans* (Macartney) Kofoid, *Karenia mikimotoi* (Miyake et Kominami ex Oda) Hansen et Moestrup, *Heterocapsa circularisquama* Horiguchi, and *Cochlodinium polykrikoides* Margalef (Dinophyceae). The top three most noxious species in the Seto Inland Sea in order of the amount of fishery damage are *C. antiqua*, *K. mikimotoi*, and *H. circularisquama* in the Seto Inland Sea.

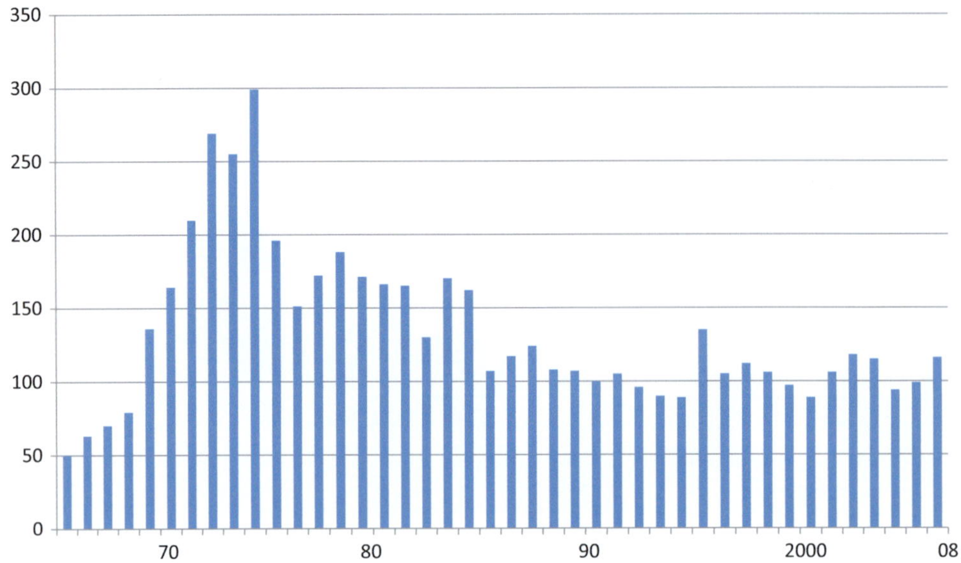

Fig. 3.24 Occurrence of red tides in the Seto Inland Sea from 1973 to 2010

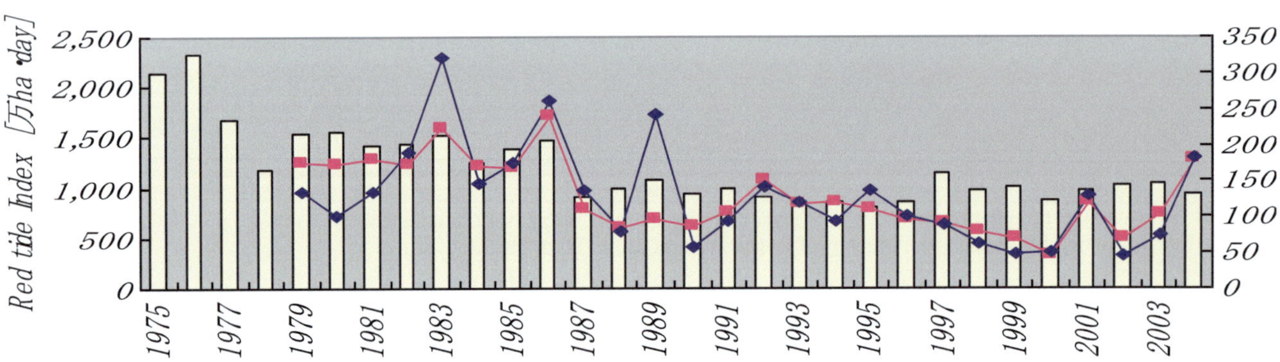

Fig. 3.25 Year-to-year variation in the Red Tide Index in the Seto Inland Sea. Square shows the result in the coastal area, diamonds the offshore area, and bars denote the number of occurrences per year (Ishii et al. 2014)

3.3.4 Hypoxia

Horizontal distributions in DO (dissolved oxygen) concentration above the sea bed in July 1981 and 2000 are shown in Figs. 3.27 and 3.28. Hypoxia (DO is lower than 3.6 mg/l) appeared in the eastern part of Osaka Bay, the central part of Harima-Nada, the eastern part of Hiuchi-Nada, the northern part of Hiroshima Bay, and the western part of Suo-Nada, where Chl.a is high (Fig. 3.16), TP and TN concentrations in sediments are high (Figs. 3.18 and 3.19), and the diversity index of macro-benthos is low (Fig. 3.23). Such distribution of hypoxia showed no change from 1981 to 2000.

3.3.5 Seagrass Beds and Tidal Flats

Seaweed and seagrass beds and the tidal flats in the Seto Inland Sea area have decreased. The former are considered to be important as a zone for nursery grounds of shells and fish. The latter plays an important role in the ecosystem and in self-purification. Trends for each total area are shown in Figs. 3.29 and 3.30.

The area of the Zostera (seagrass) zone is 6,381 ha, while the Garamo (Sargasso) zone is 5,511 ha, and the Ulva and Enteromorpha zone is 4,667 ha. As for the area of the tidal flats, the largest one, covering an area of 6,409 ha, is located in the western part of Suo-Nada and the second largest, 1,022 ha, is in Hiuchi-Nada.

Some fishermen in the Seto Inland Sea have rehabilitated the seagrass beds of eelgrass (Zostera) by spraying their seed for more than 30 years and have succeed in recovering about one third of the area compared to the maximum area of seagrass beds 60 years ago (Yanagi 2012).

3.3.6 Fish Catch

The Seto Inland Sea has an extremely high productivity per unit area of fishery product, which may be due to the two following reasons: one is that essential nutrients needed for

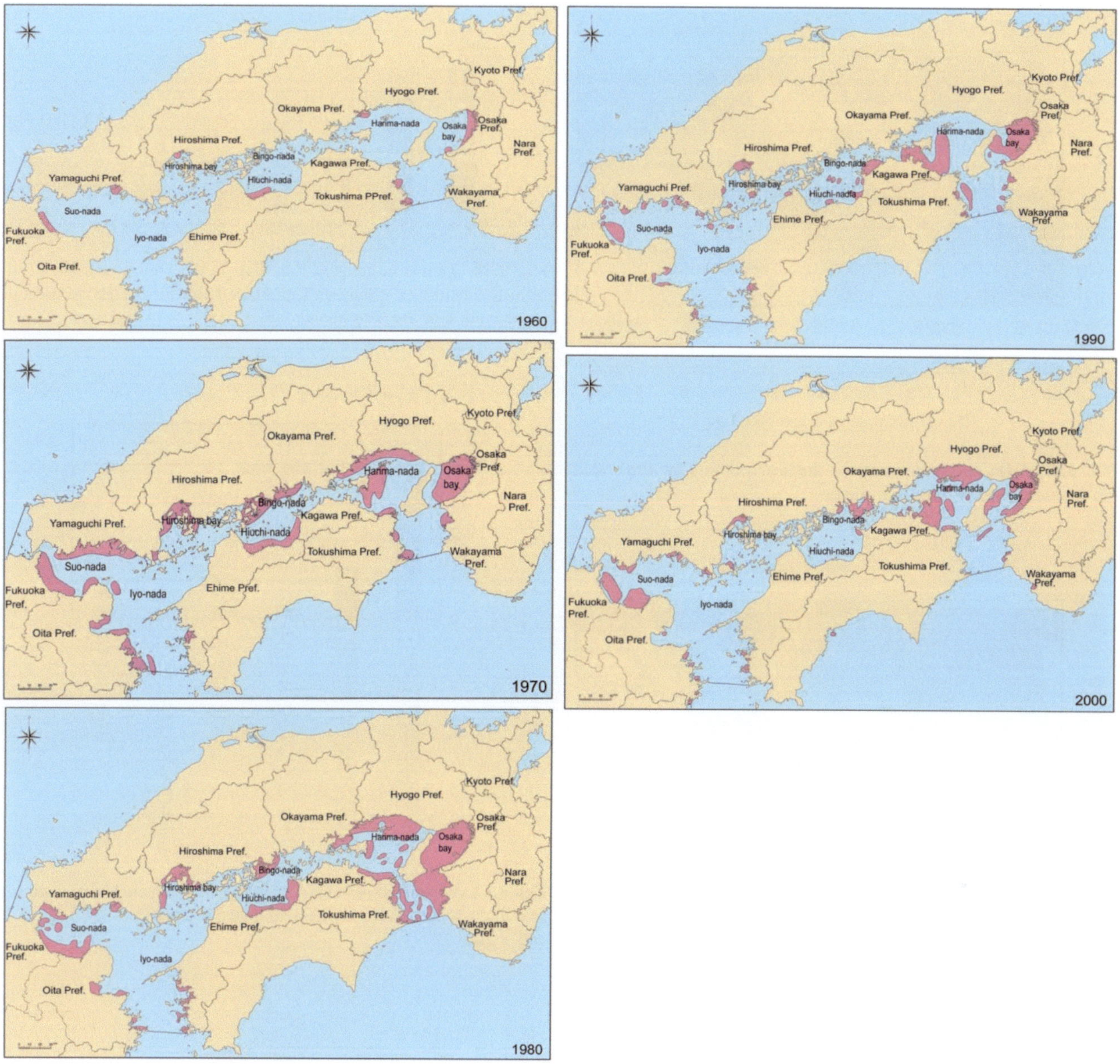

Fig. 3.26 Decadal changes in the distribution of red tide in the Seto Inland Sea

the growth of fish and shells are available thanks to having a sufficient discharge from the many rivers around the sea. The other is the extremely complex and semi-enclosed geography of the sea with its numerous islands, straits, nadas, and the various sizes of its bays (Fig. 3.31, Okaichi and Yanagi 1997). Fish species and the compositions of fisheries and aquaculture productions in 2005 are shown in Figs. 3.32 and 3.33, respectively.

In Fig. 3.32 the total amount of fishery production in the Seto Inland Sea was 198,000 t with the ratio of anchovy (18 %), white bait (10 %), and sand eel (10 %) being rather large. In addition, Fig. 3.32 shows that the total amount of aquaculture product in the Seto Inland Sea was 286,000 t with its large ratio of oysters (47 %) and "nori" (laver; seaweed) (42 %), occupying nearly 90 % of the total.

Oyster and nori culture are very popular in the Seto Inland Sea, which enjoys very calm sea weather as shown in Fig. 3.33.

Figure 3.34 shows the change in fishery production, which increased until around 1985, but it has since been decreasing. The maximum fish catch in the Seto Inland Sea was about 460,000 t/year in 1985. The reasons for this decrease in fish catch, other than anchovy, which has a dominant variability with a period of about 50–70 years on a global scale, are discussed in Sects. 3.3.4 and 3.3.5. On the other hand, average TP and TN concentrations in the Seto

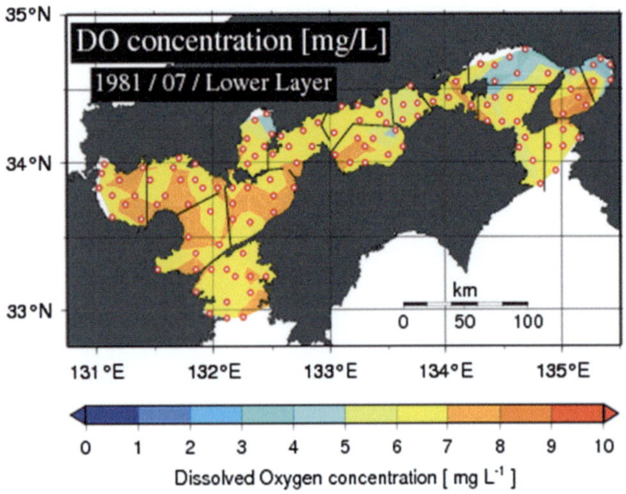

Fig. 3.27 DO concentration above the sea bed in July 1981. *Red circles* indicate the observation station

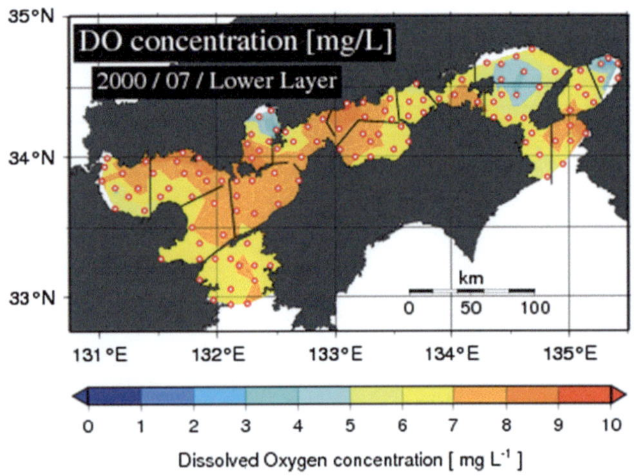

Fig. 3.28 DO concentration above the sea bed in July 2000. *Red circles* indicate the observation station

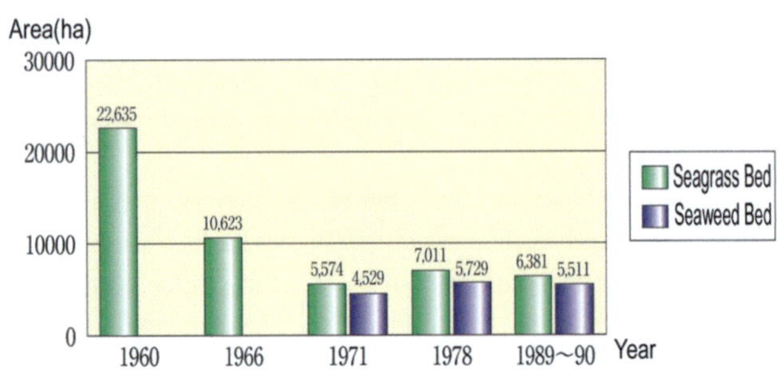

Fig. 3.29 Trends in areas of seaweed and seagrass beds in the Seto Inland Sea (Data from Setouchi-Net)

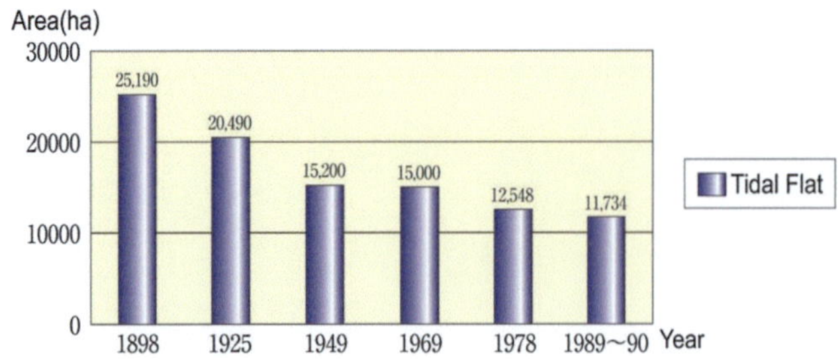

Fig. 3.30 Trends in areas of tidal flats in the Seto Inland Sea (Data from Setouchi-Net) (Source: The Association for the Environmental Conservation of the Seto inland Sea)

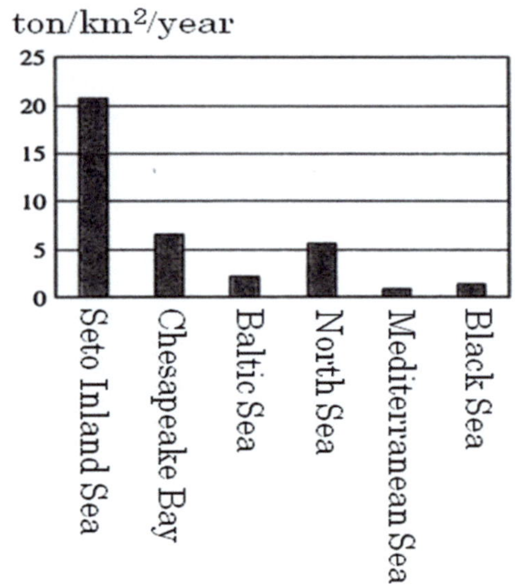

Fig. 3.31 Fish catch per unit area per year in selected semi-enclosed seas in the world. Value for the Seto Inland Sea based on 1985 (Okaichi and Yanagi 1997)

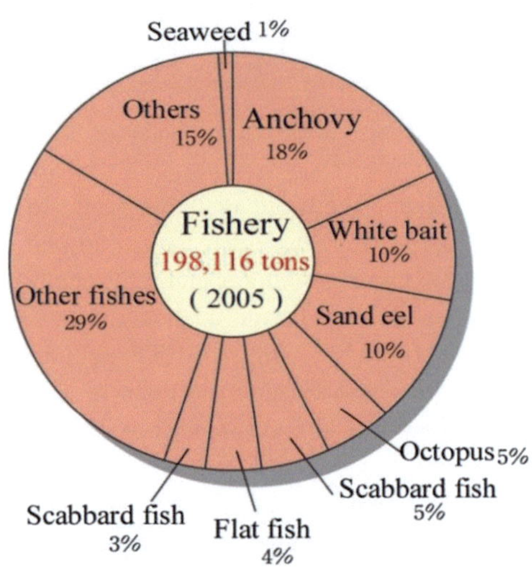

Fig. 3.32 Fish species and composition of fishery production (Source: Ministry of Agriculture, Forestry and Fisheries)

Inland Sea did not change and kept nearly the same value from 1980 to 2005. It is therefore believed that the decrease in fish catch does not have a direct relation to concentrations of TP and TN in the Seto Inland Sea. However, averaged TP and TN concentrations began to decrease from 2007 and may relate to oligotrophication which will be discussed in Sect. 3.3.4.

Various kinds of aquaculture projects have been undertaken to compensate for the deterioration in the living environment of fish and shells, which have a positive effect on the marine production industry in the Seto Inland Sea.

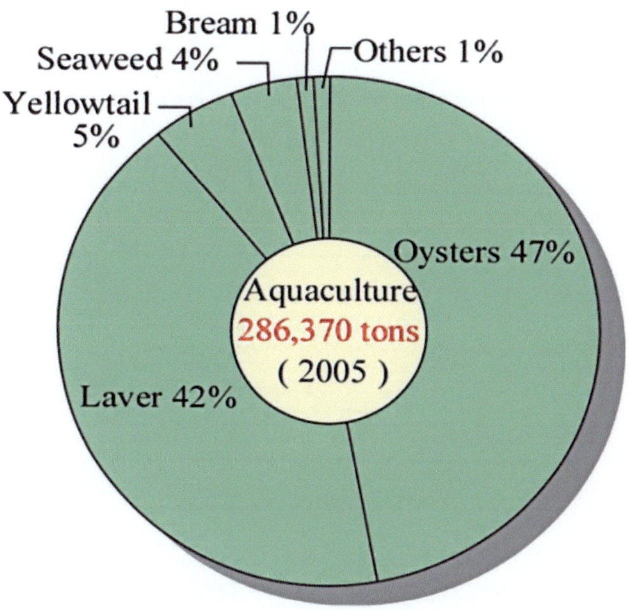

Fig. 3.33 Fish species and composition of aquaculture production (Source: Ministry of Agriculture, Forestry and Fisheries)

3.4 Responses

3.4.1 Special Law

The Tentative Law Concerning Special Measures for Conservation of the Environment of the Seto Inland Sea (Interim Law) was established on October 2, 1973, based on which the Permanent Law was established on June 13, 1978. The major objective of this Law was to decrease TP and TN loads from land flowing into the Seto Inland Sea in order to stop eutrophication there.

3.4.2 Loads from Land

Trends in TP load are shown in Fig. 3.35. The TP load has tended to decrease slowly since 1974 when the guidance for a reduction in phosphorus load was introduced under the framework of the Special Law. TN load showed a tendency to increase from 1984 as shown in Fig. 3.36. Accordingly, guidance for reduction of TN load was started in 1996 under the framework of the Special Law, and the TN load has decreased since 1994 as shown in Fig. 3.36.

The reduction in TP and TN loads has resulted in the decrease in the occurrence of red tides in the Seto Inland Sea as shown in Fig. 3.24, but it has not resulted in improvement

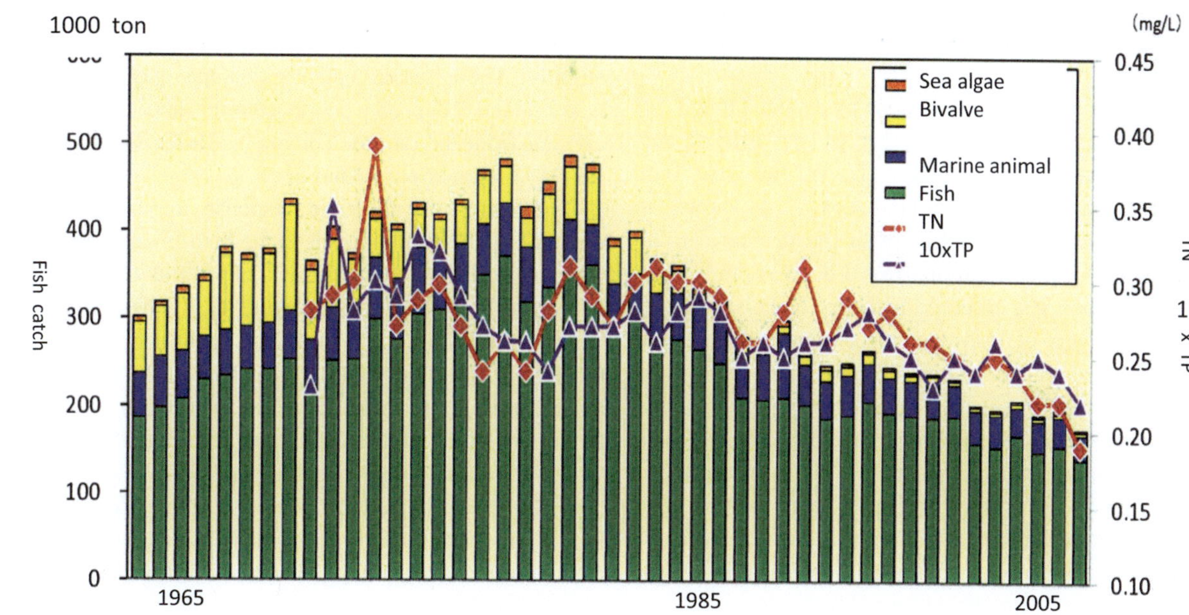

Fig. 3.34 Trends in fishery production and averaged TP and TN concentrations in the Seto Inland Sea (Source: Ministry of Environment)

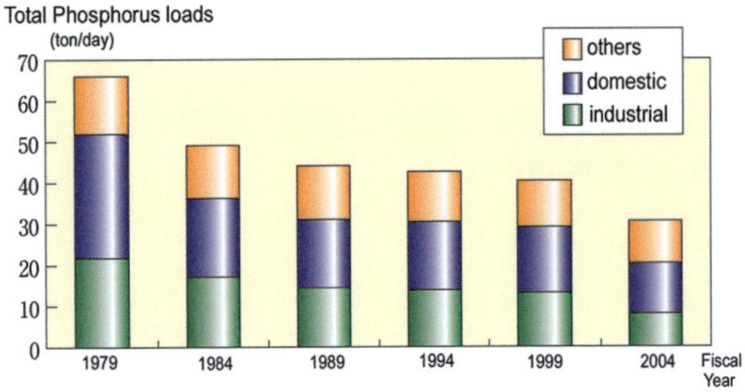

Fig. 3.35 Changes in the total amount of phosphorus load in the Seto Inland Sea (Data from Setouchi-Net) (Source: Ministry of Environment of Japan)

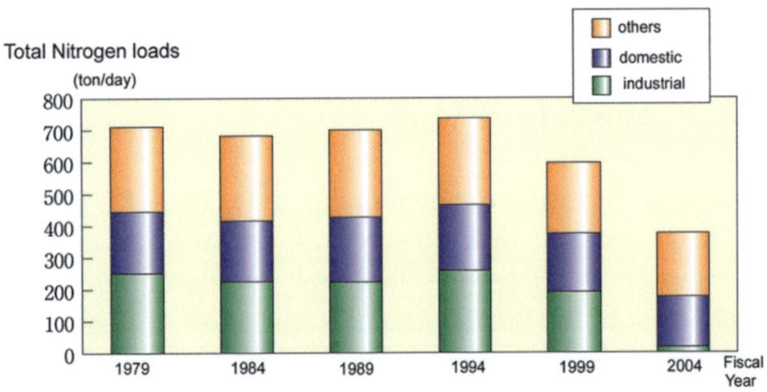

Fig. 3.36 Changes in the total amount of nitrogen load in the Seto Inland Sea (Data from Setouchi-Net) (Source: Ministry of Environmental of Japan)

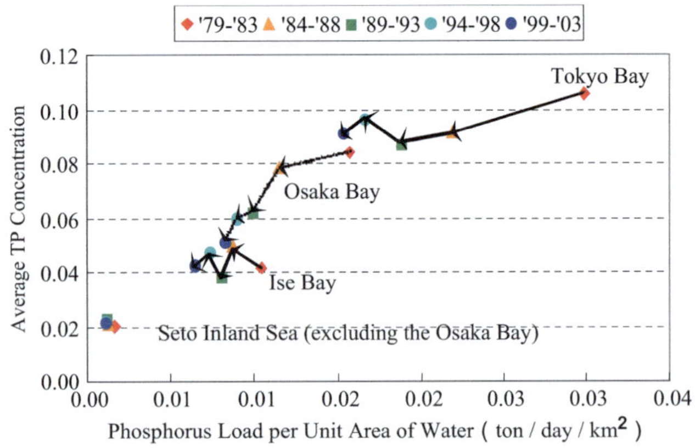

Fig. 3.37 Relationship between TP load and average TP concentration in surface sea water (Source: Ministry of Environment of Japan (2005))

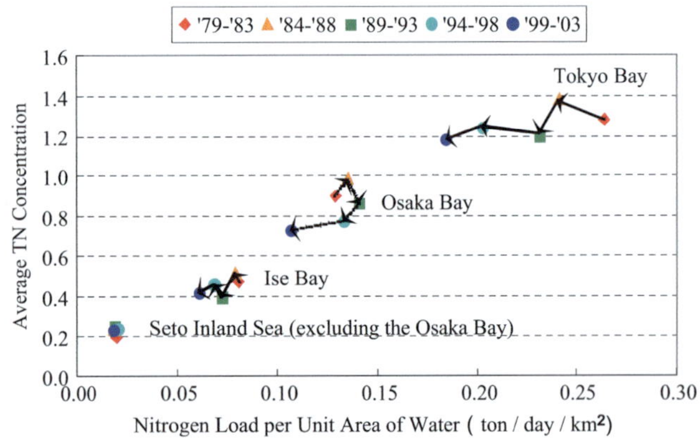

Fig. 3.38 Relationship between TN load and average TN concentration in surface sea water (Source: Ministry of Environment of Japan (2005))

in water quality (TP and TN concentrations) other than in Osaka Bay.

3.4.3 Response in Water Quality and Contribution of Load from Ocean

The relationship between TP and TN loads and TP and TN concentrations in the surface water of the four largest enclosed coastal seas in Japan – Tokyo Bay, Ise Bay, Osaka Bay, and the Seto Inland Sea – is shown in Figs. 3.37 and 3.38. Correlating with the decrease in TP and TN loads, a generally proportional decrease in TP and TN concentrations was observed in Tokyo and Osaka Bays, although the levels of load and concentration differ area-specifically. It should be noted that the nutrient level of Osaka Bay in the Seto Inland Sea is very high compared with that in other areas of the Seto Inland Sea. Estimated TP load in 1979 amounted to 62.91 t/day in the Seto Inland Sea and 41.2 t/day in Tokyo Bay, while TN load was 666 t/day and 364 t/day, respectively. According to the decrease in TP and TN loads, relatively proportional decreases in TP and TN concentrations were also observed in Tokyo and Osaka Bays, but not Ise Bay and the Seto Inland Sea, although the relationships between phosphorus and nitrogen differ slightly. These results suggest that the quality of surface water mainly depends on phosphorus and nitrogen loads coming from the land in Tokyo and Osaka Bays but not in Ise Bay and the Seto Inland Sea.

The reason for different response in water quality to decrease of TP and TN loads in the Seto Inland Sea (excluding Osaka Bay) from that in Tokyo and Osaka Bays is explained by Yanagi and Tanaka (2013), where it is pointed out that the origin of TN and TP is mainly from the Pacific Ocean and the bottom as shown in Fig. 3.39. As shown, about 56–57 % of TP and TN in the Seto Inland Sea originate from the Pacific Ocean, and 33 % of TP and 28 % of TN from the bottom by release. Only 11 % of TP and 15 % of TN originate from the land. Therefore, the effect of decreasing TP and TN loads from the land does not have a large effect on the water quality in the Seto Inland Sea. The ratio of airborne TP and TN to the Seto Inland Sea is less than 10 % (Yanagi 1997).

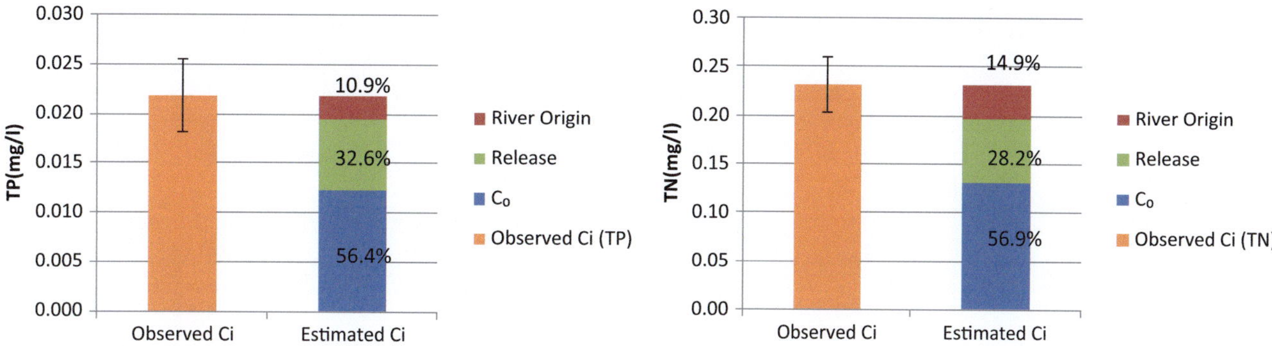

Fig. 3.39 The origin of TP and TN in the Seto Inland Sea (Yanagi and Tanaka 2013)

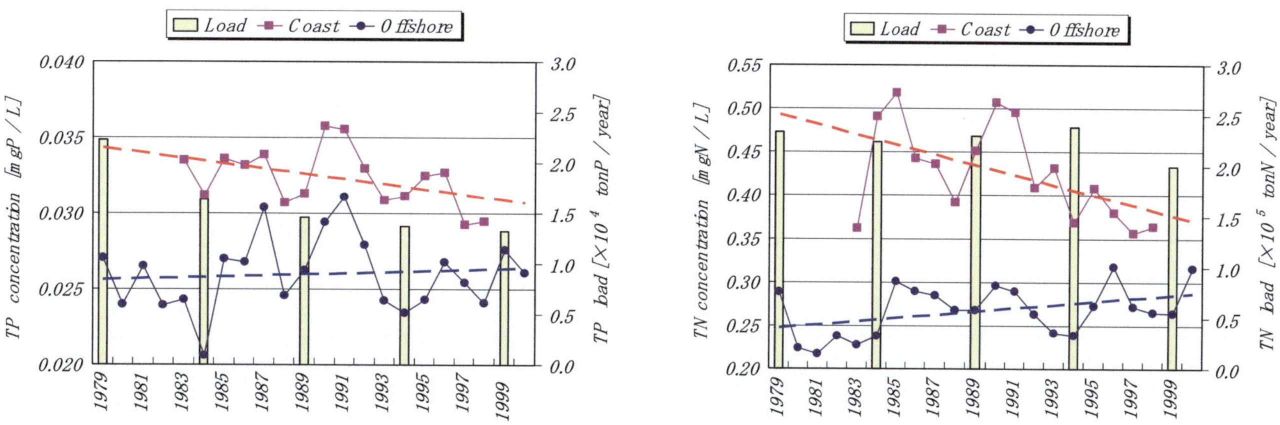

Fig. 3.40 Variation in TP and TN concentrations in the nearshore and offshore areas of the Seto Inland Sea (Ishii and Yanagi 2005)

Year-to-year variations in TP and TN concentrations in the Seto Inland Sea are shown in Fig. 3.34. TP and TN concentrations did not change during these 20 years as shown, except for the last 4 years. Although average TP and TN concentrations in the offshore area of the Seto Inland Sea did not change or increased a little during this time, concentrations in the coastal area within 2 km of the coast decreased as shown in Fig. 3.40 (Ishii and Yanagi 2005).

3.4.4 Reclamation

Heavy industries such as steel, petrochemicals, and shipbuilding have been concentrated in the Seto Inland Sea since the 1950s. In the process, many coastal areas have been reclaimed for industrial sites and ports. The total area of permitted reclamation is very distinct from 1965 to 1973, but decreases drastically from 1974 as shown in Fig. 3.41. It was in this year that the "Tentative Law Concerning Special Measures for Conservation of the Environment of the Seto Inland Sea" was established. However, reclamation continued in the Seto Inland Sea after 1974 because this Special Law permits the reclamation for public use only, e.g. reclamation for the New Kansai Airport in 1987 or reclamation for waste dumping in 1999.

The land area reclaimed in the Seto Inland Sea since 1898 amounts to 455 km^2, which is equivalent to about 70 % of the area of Awaji Island (the largest island in the Seto Inland Sea). Such large-scale reclamation means that about 20 % of shallow sea with a depth less than 10 m has been reclaimed and has dramatically destroyed seagrass and seaweed beds, tidal flats (as shown in Figs. 3.29 and 3.30), and marine life including benthos.

3.4.5 Oligotrophication

Yamamoto (2003) was the first to point out that the Seto Inland Sea began to undergo oligotrophication based on the decrease of fish catch data from the mid-1980s to the mid-1990s. It does not fully describe the actual situation if we say such a decrease in the fish catch resulted from oligotrophication, since TP and TN concentrations did not decrease in the Seto Inland Sea during the same period as shown in Fig. 3.33.

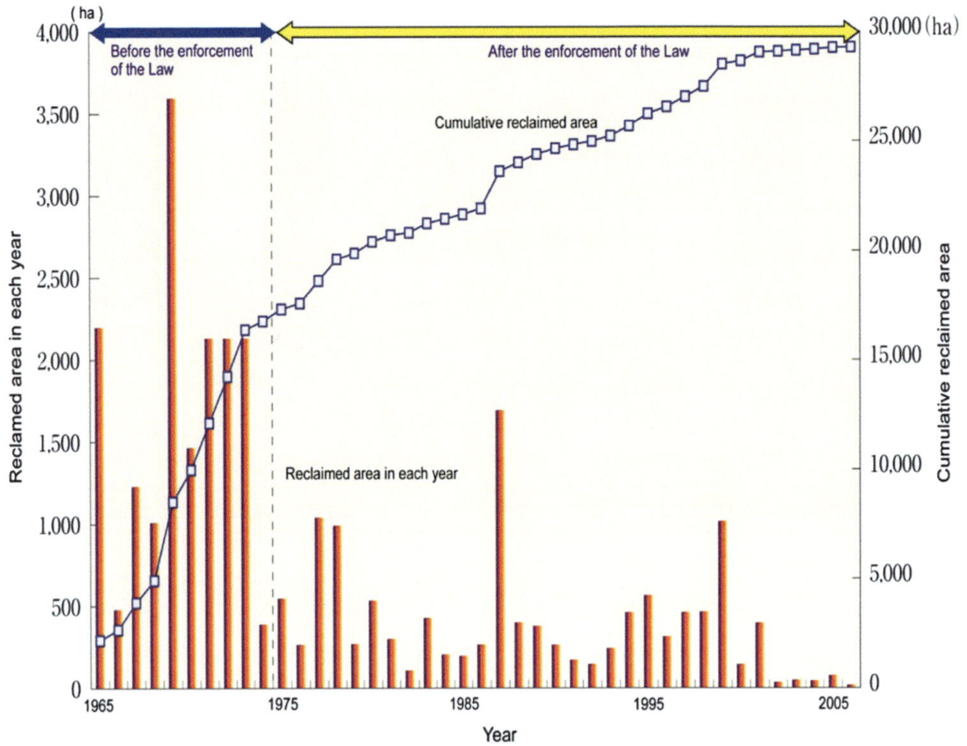

Fig. 3.41 Trends in reclaimed areas in the Seto Inland Sea (Source: Ministry of the Environment of Japan; Data for 1971-1973 show the total from January, 1971 to November 1, 1973; Data after 1973 show the total from November 2 of the previous year to November 1, the following year; A tentative law was enforced on November 2, 1973; Values for 1971–1973 are mean values for the 3 years)

However, the situation has changed in recent years in the Seto Inland Sea. DIN (dissolved inorganic nitrogen) concentration in Harima-Nada and the Bisan Strait has decreased since 1994 (Tada et al. 2010) though TN concentration has remained at nearly the same value. The DIN value in Harima-Nada is only about one tenth of TN (Fig. 3.42). Thus, the limiting nutrient for primary production in Harima-Nada has changed from DIP (dissolved inorganic phosphorus) to DIN from that time as shown in Fig. 3.43. The cell density of the diatom, which is the main species of phytoplankton in Harima-Nada, showed no change but the major species of diatoms has changed from *Skeletonema* spp. to *Chaetoceros* spp. (Fig. 3.44).

Fishing by trawling net began to decrease 2 years after the concentration of DIN decreased in Harima-Nada during the winter as shown in Fig. 3.45 (Tanda and Harada 2011). Nori (sea laver) production in the Bisan Strait has also decreased in recent years as shown in Fig. 3.46 (Tada et al. 2010).

Only a change in DIN without that of TN (Fig. 3.42) suggests that the decomposition of TN into DIN decreases; in other words, material cycling of nitrogen by biochemical processes was active in the past but is not active now. It is suggested that the material cycling of biochemical elements including DIP, DOP (dissolved organic phosphorus), POP (particulate organic phosphorus), DIN, DON (dissolved

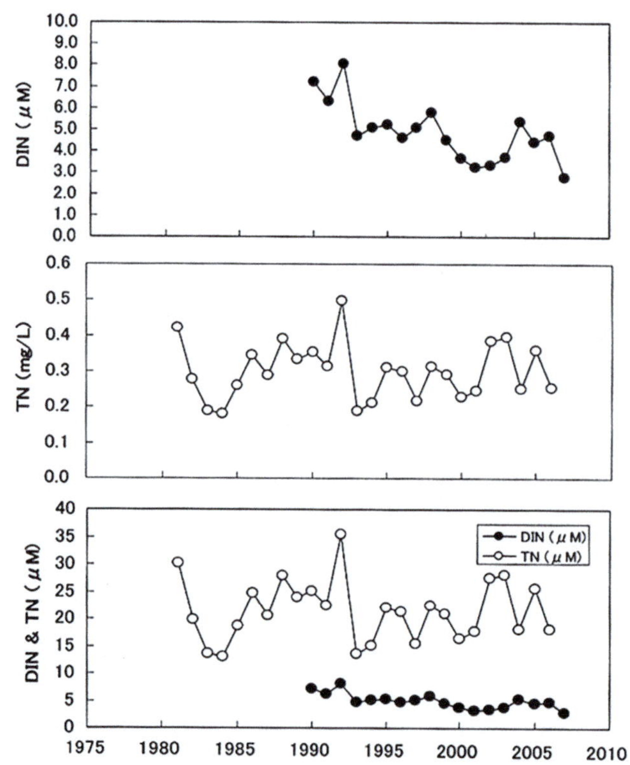

Fig. 3.42 Year-to-year variations in DIN and TN at Harima-Nada in Kagawa Prefecture (Tada et al. 2010)

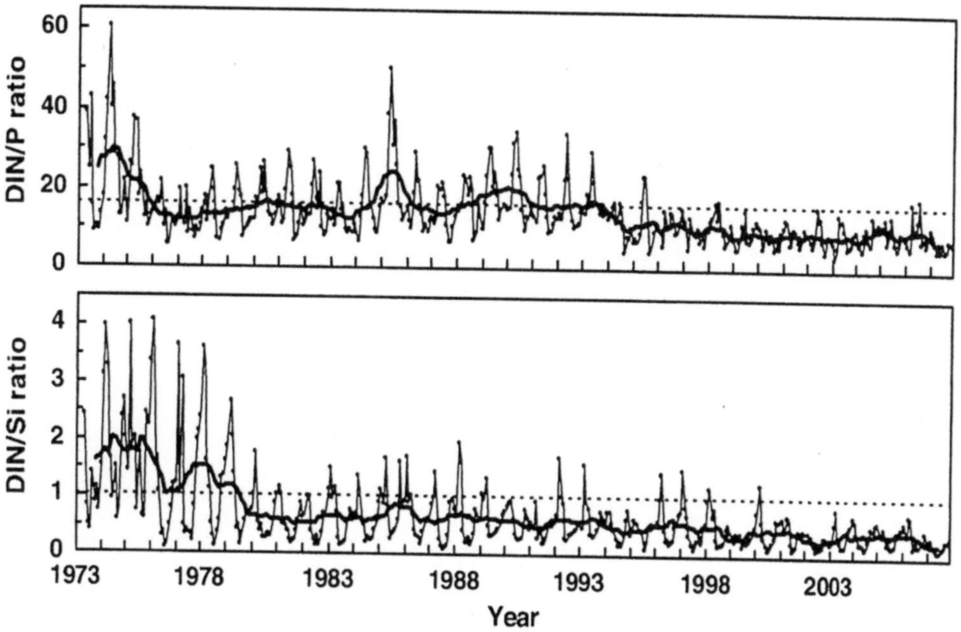

Fig. 3.43 Long-term variations in DIN/DIP and DIN/DSi molar ratios in Harima-Nada over the 35 years from April 1973 to December 2007. Monthly data were averaged for three depths at 19 sampling stations. The smoothed lines were derived from a 13-month moving average. The dashed horizontal lines represent the Redfield ratio (Nishikawa et al. 2010)

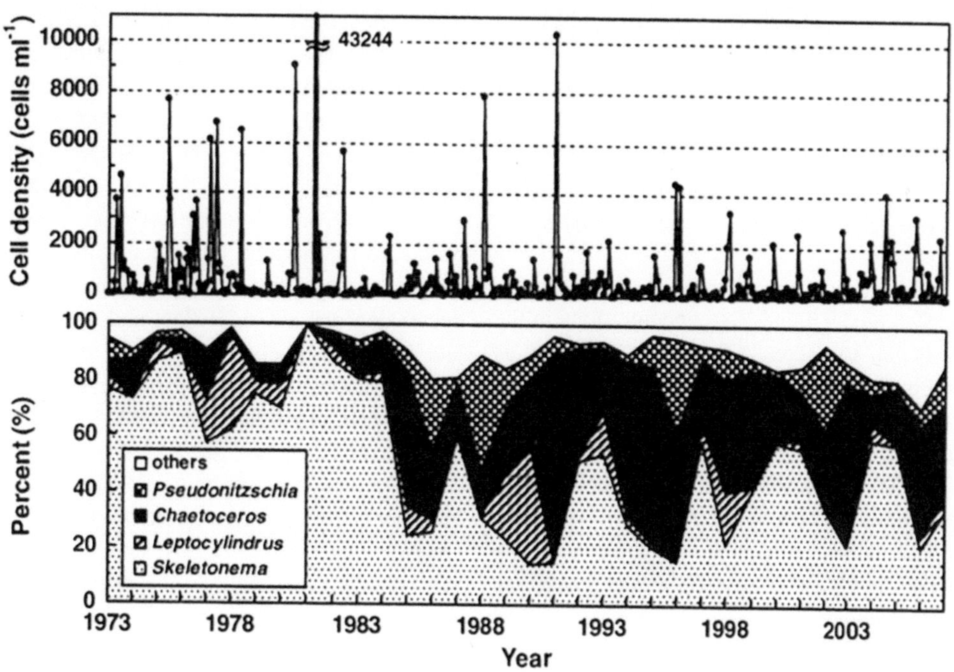

Fig. 3.44 Long-term variations in monthly total cell density and yearly percent species composition of diatoms in the surface layer of Harima-Nada from April 1973 to December 2007. Data on monthly cell densities and yearly composition of species are the average of monthly sampling at 19 stations (Nishikawa et al. 2010)

organic nitrogen), PON (particulate organic nitrogen) has changed in the marine environment, especially the change in related marine ecosystem.

We have to revert to a suitable marine environment based on a healthy marine ecosystem in order to recover healthy and comprehensive material cycling in the Seto Inland Sea.

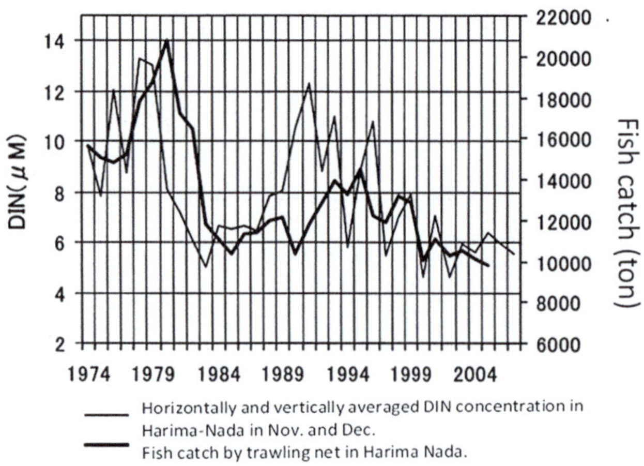

Fig. 3.45 Year-to-year variations in DIN concentration in Harima-Nada and fish catch by trawling net in Harima-Nada (Tanda and Harada 2011)

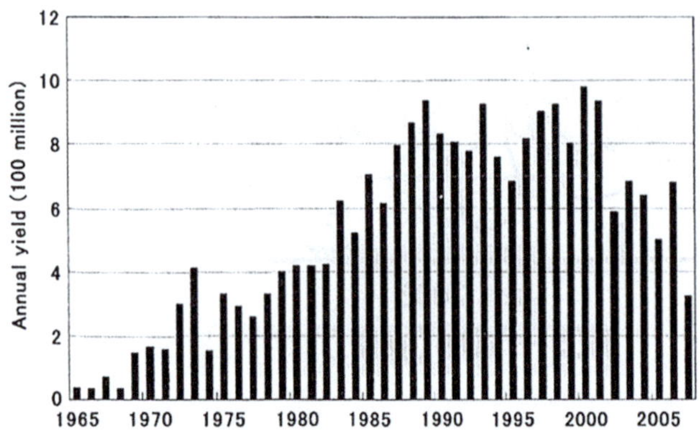

Fig. 3.46 Year-to-year variations in seaweed culture production in Bisan Strait (Tada et al. 2010)

However, the averaged TP and TN concentrations in the Seto Inland Sea began to decrease from 2009 and this is thought to be due to the total TP and TN load control from the land under the framework of the Special Law and also due to the decrease in the bottom release of TP and TN as suggested from Fig. 3.39 as the TP and TN concentrations in the Pacific Ocean have not changed. Therefore, oligotrophication will continue in the future due to the coupling effect of a decrease in TP and TN loads and the decrease in mineralization by biochemical processes.

3.4.6 History of Eutrophication and Oligotrophication

The history of eutrophication and oligotrophication in the Seto Inland Sea is schematically shown in Fig. 3.47. In the 1960s, when the period of rapid economic growth began,

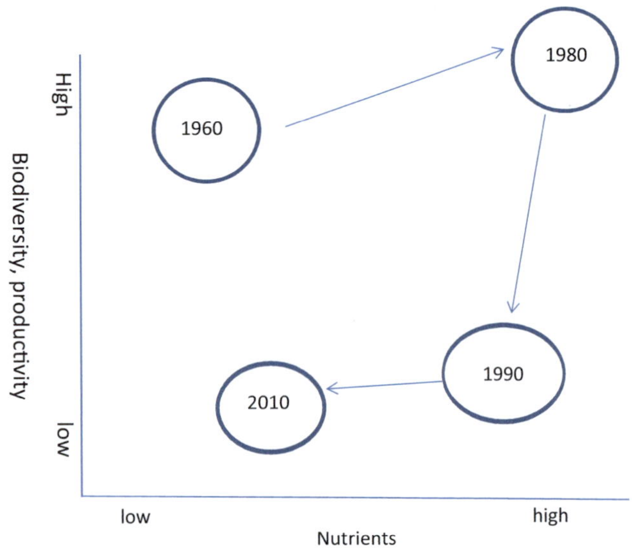

Fig. 3.47 History of eutrophication and oligotrophication in the Seto Inland Sea

nutrient concentrations were low but biodiversity was high, though fish production was not low in the Seto Inland Sea. Due to the increase in nutrient loads from the land during the period of rapid economic growth, nutrient concentrations increased and fish production also increased. The maximum fish catch was attained in 1985. Control of total the nutrient load was begun in 1979 under the framework of the Special Law but the concentrations of TP and TN in the Seto Inland Sea did not change until the 1990s due to the large supply of nutrients from the Pacific Ocean compared to the amount coming from the land. However, biodiversity and the fish production continued to decrease due to hypoxia and the destruction of shallow areas environment, especially the decreasing area of tidal flats and seagrass beds. From the late 1990s, DIN began to decrease mainly due to the change in material cycling of nitrogen related to the change in biochemical processes that resulted from changes in marine ecosystem.

Our main objective is to clarify how best to return the biodiversity and production in the Seto Inland Sea from the recent state (2010) to its past state (1960). It is thought that the pathway of changing nutrient concentrations and biodiversity or production during eutrophication is different from that during oligotrophication; in other words, a multi-phase steady state may exist under the same nutrient concentration as shown in Fig. 3.48. Biodiversity and production during eutrophication are thought to be higher than during oligotrophication at the same concentration of nutrients due to the clockwise hysteresis resulting from the effects of hypoxia. We have to clarify such a process quantitatively.

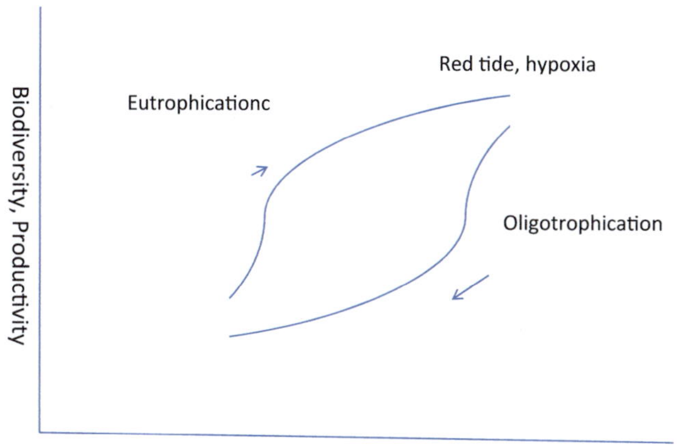

Fig. 3.48 Multiphase quasi-steady state of material cycling and nutrient concentration

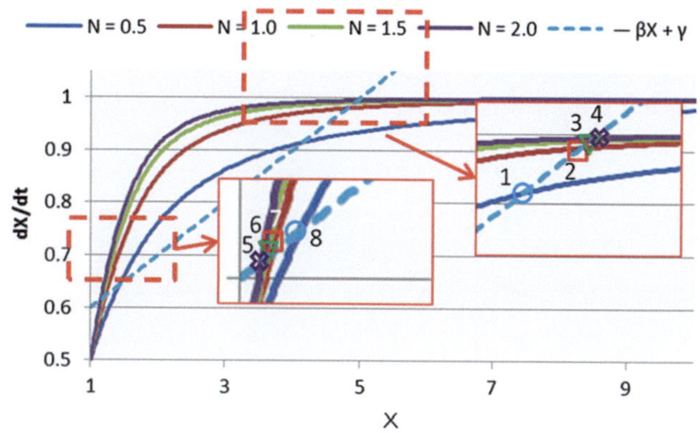

Fig. 3.49 Relation between phytoplankton growth and death and grazing rates when $dX/dt = 0$

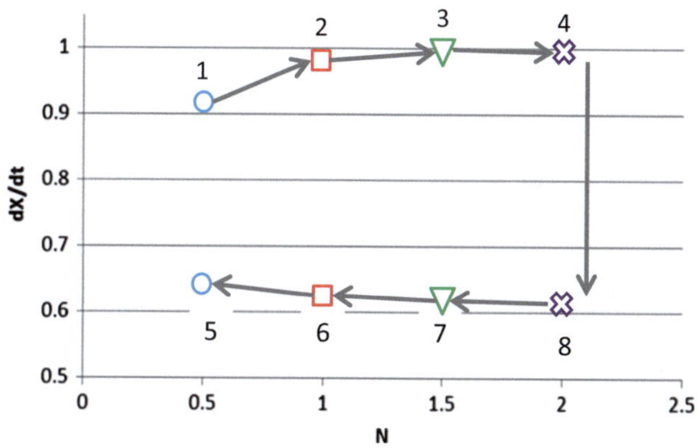

Fig. 3.50 Relationship between nutrient concentration and phytoplankton community biomass

3.4.7 Dynamics of Eutrophication and Oligotrophication

The temporal change in biodiversity and productivity is very difficult to study theoretically. Therefore, we look at the temporal change in the phytoplankton community biomass and its relation to the nutrient concentration.

The temporal change in the cell density of phytoplankton (X') is defined by the following Michaelis-Menten formula:

$$\frac{dX'}{dt} = G\frac{N}{N + K_S}X' = \alpha X' \quad (3.1)$$

where G, N, and Ks express the maximum growth rate, nutrient concentration, and half-saturation constant, respectively. Therefore, X' is as follows:

$$X' = e^{\alpha t} \quad (3.2)$$

On the other hand, the temporal change in the phytoplankton community biomass (X) is defined by the following sigmoid (logistic) function:

$$\frac{dX}{dt} = \frac{cX^p}{X^p + h^p} \quad (3.3)$$

where c, p, and h express the constant numbers and $X = \delta X'$. When we consider not only growth but also grazing (beta) and death (γ), Eq. (3.3) is changed to

$$\frac{dX}{dt} = \frac{X^p}{X^p + 1} - (\beta X + \gamma) \quad (3.4)$$

Equation (3.4) takes the quasi-steady state ($dX/dt = 0$), when the growth rate (first term in the right-hand side) equals the death and grazing rate (second term in the right-hand side). Such a situation is shown in Fig. 3.49 under the condition of $p = 1$, beta $= 0.1$, $\gamma = 0.5$, and $\delta = 1$ with different nutrient concentrations of N.

The relation between nutrient concentration and phytoplankton community biomass is shown in Fig. 3.50.

Figure 3.50 corresponds well with Fig. 3.48 qualitatively. The transition from point 4 to point 8 is called the "regime shift" and this regime shift is thought to happen due to the effect of hypoxia. Another regime shift from points 5 or 6 to points 1 or 2 is expected to happen as a result of the disappearance of hypoxia in the Seto Inland Sea.

The reason for the different response in fish catch to the same nutrient concentration is explained by Fig. 3.51. At point A during eutrophication, the dead phytoplankton after a red tide become good bate for the benthos and bottom fish, but at point B, during oligotrophication the dead phytoplankton after a red tide increase the degree of hypoxia due to consumption of oxygen for decomposition. The difference between A and B is based on the accumulation of organic matter in the bottom sediment during eutrophication.

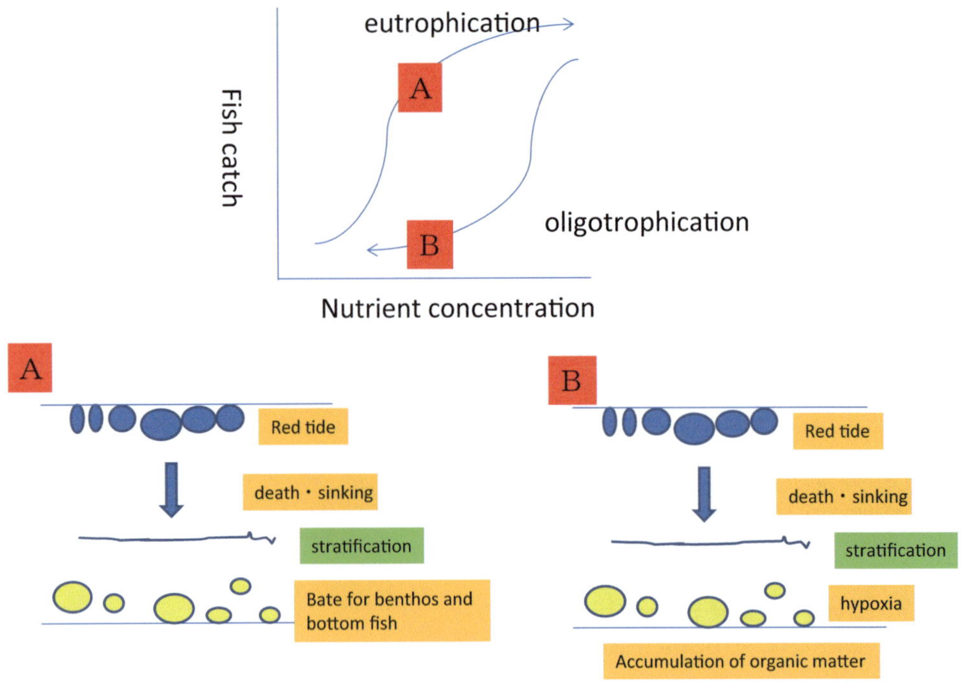

Fig. 3.51 Difference between A (eutrophication period) and B (oligotrophication period) during summer in the Seto Inland Sea

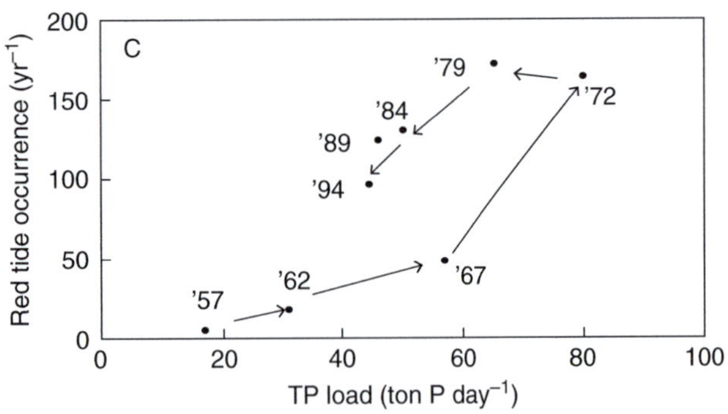

Fig. 3.52 Relationship between TP load variation and the number of red tide occurrences per year (Yamamoto 2003)

A similar hysteresis to Fig. 3.48 is shown in the relationship between the variations in TP load and red tide occurrence as shown in Fig. 3.52 (Yamamoto 2003). In this case the hysteresis is counterclockwise and is also explained by Fig. 3.51, that is, the TP concentration during oligotrophication is higher than during eutrophication due to the DIP release from the bottom sediment at the same TP load.

3.5 Future Tasks

3.5.1 New Environmental Policies

New environmental policies contributing to the recovery of a sound environment were officially introduced to the Seto Inland Sea in 2000 (when the fifth water quality control policy was confirmed). Based on previous environmental conditions, the main target of the policy was changed from water quality control to environmental remediation and restoration of habitat. Led by the new policy for the Seto Inland Sea, a new law for the promotion of nature was enacted in 2002. This new law was applied not only to the Seto Inland Sea but throughout Japan. Collaboration among various groups, such as local and national governments, local residents, NGOs, not-for-profit organizations (NPOs), scientists, and fishermen, is expected to play an important role in promoting individual restoration projects.

Furthermore, in 2006, the sixth water quality control policy was decided on, according to which the total load control policy for TP and TN was maintained only in Osaka Bay but abolished in other bays and nadas of the Seto Inland Sea.

The Basic Act on Ocean Policy was enacted in July 2007, under which the new Headquarters for Ocean Policy takes the responsibility in all governmental activities related to marine affairs and marine environment.

Four possible causes for the decrease in fish stocks in the Seto Inland Sea have been proposed, namely:

1. Changes in the natural environment due to a regime shift (large-scale climatic and oceanographic changes)
2. Overfishing
3. Destruction of spawning and nursery grounds or habitats
4. Long-term changes in the ecosystem due to the effect of human activities

Decreases in sand eel stocks were more directly affected by habitat destruction due to large-scale sea-sand mining for the concrete and construction industries, although sea-sand mining in the Seto Inland Sea was prohibited in 2006.

Among the four possible causes identified above, countermeasures against regime shift and long-term ecosystem change are very difficult or almost impossible to be achieved in a relatively short period of time. Realistic countermeasures can only be taken against overfishing and against destruction of spawning and nursery grounds or habitats. This is where the importance of habitat restoration is key – in particular of tidal flats, seaweed bed, and seagrass bed – and of living resource management in shallow areas.

Shallow coastal waters, seaweed, and seagrass beds are important habitats and reproduction grounds for many marine organisms. Tidal flats provide the most important reproduction area for bivalves. They also play an important role in the decomposition of organic matter, e.g. from DON and PON to DIN. Both of these important habitats have been undergoing long-term decreases in the Seto Inland Sea area. Between 1978 and 1991, 1,500 ha of seaweed and seagrass beds and 800 ha of tidal flats were lost from the Seto Inland Sea as shown in Figs. 3.29 and 3.30, mainly due to reclamation, dredging, or other human activities. As a result, a significant portion of natural coastline has been converted to man-made coastline, consisting of upright concrete structures as shown in Fig. 3.10. These have not provided a good habitat for many organisms living along the seashore, nor have they provided the valuable functions of a natural coastline such as purification of organic pollution and de-nitrification. Such decreases in seaweed and seagrass beds and tidal flats means that the material cycling changed from multi-paths to a simple path as shown in Fig. 3.53.

Of the many enclosed coastal seas of Japan, the Seto Inland Sea is one of the main sites for environmental remediation and restoration. Remediation and restoration carried out by different organizations in the Seto Inland Sea cover a variety of methods, depending on their objectives. Some examples include the simple restoration of tidal flats or seaweed and seagrass beds or a combination of tidal flats and seaweed and seagrass beds, artificial rocky shores, artificial lagoons, artificial submerged slopes, beneficial use of dredged sediments, and bird sanctuaries. Some typical examples of such activities are described below.

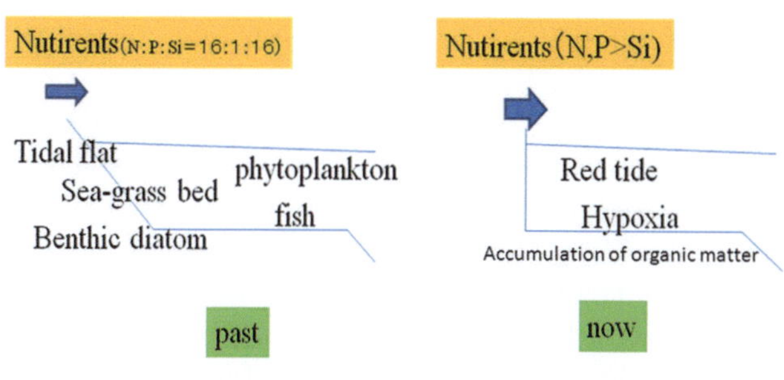

Fig. 3.53 Material cycling in the past (*left*) and present (*right*) in the Seto Inland Sea. Nutrient loads from the land increased with the change in N:P:Si ratio. Nutrients loads from the land were assimilated to benthic diatom, seaweed and seagrass, as well as the phytoplankton in the past but they were mainly assimilated to phytoplankton, and red tide and hypoxia or anoxia occurs now. Furthermore, the nutrient loads have begun to decrease in recent years

1. Very comprehensive habitat restoration is being conducted at the estuary of the Fushino River in Yamaguchi Prefecture, including the environmental remediation of the Fushino River watershed. Reforestation of the upstream area is also included in this project. A local currency, called Fushino, was introduced to the area in order to promote the project which was supported by the wide variety of stakeholders.
2. Etashima Bay in Hiroshima Prefecture has a highly enclosed topography and is an important ground for the culture of oysters. Oxygen depletion in the bottom water in summer and deterioration in sediment quality have been serious problems there. The local government of Hiroshima Prefecture initiated the restoration of Etashima Bay using a multi-sectoral approach in which five prefectural research institutes participated in the development of efficient tidal flats and seagrass beds in order to activate local fisheries and oyster culture.
3. Along the coast of Kansai International Airport, which is located on an artificial island in Osaka Bay, a gentle slope of natural rocks and stones rather than a vertically uplifted concrete wall was used in the construction of the airport. This environmentally friendly, gentle slope provided an appropriate site for seaweed beds and a suitable habitat for many kinds of organisms. As a result, this artificial structure is now working as a new seaweed bed and habitat. As the original site on which the airport was constructed was an area of muddy sea bottom, this is a good example of environmentally friendly, creative regeneration of the environment. Although the effect of the artificial island should be evaluated correctly, the effect of a more natural gentle slope itself should also be evaluated since the newly created seaweed bed plays a mitigating role in the widely lost seaweed bed in Osaka Bay.
4. At Mizushima-Nada in Okayama Prefecture, the water discharged from the dam is increased during summer when the precipitation decreases in order to increase material cycling in the coastal water because the decrease in river discharge results in a decrease in estuarine circulation and stagnant coastal water as shown in Fig. 3.54.

The above examples are described in detail in Yanagi (2012)

As to restoration of large-scale sea-sand mining effects, the environmental change is being monitored and the basic design for a restoration plan has been discussed by the Ministry of Land, Infrastructure, and Transport and the Fisheries Agency (http://seto-eicweb.pa.cgr.mlit.go.jp/rest/index.html), but practical restoration activities have not yet been achieved.

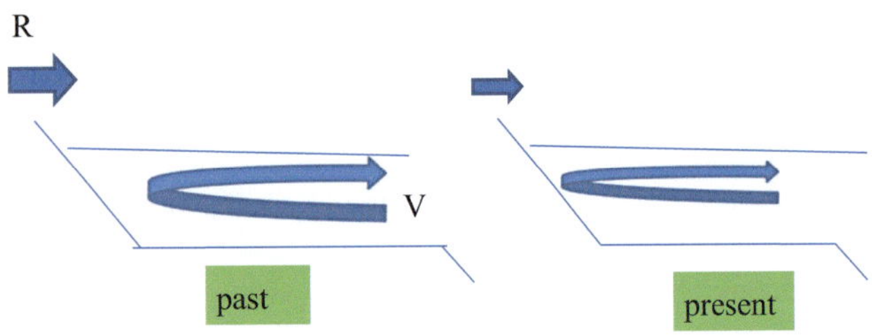

Fig. 3.54 River discharge and estuarine circulation in the past (*left*) and present (*right*) in the Seto Inland Sea. River discharge (R) has decreased due to dam construction and the volume of estuarine circulation (V) with a decrease in volume of R of about ten times; this may result in the occurrence of red tide and hypoxia

3.5.2 Future Direction

The results of the preliminary examination of the environmental health of the Seto Inland Sea made clear the deterioration in habitat conditions and ecosystem services. Hence, the restoration of such habitats in shallow water, including tidal flats and seaweed and seagrass beds, is one of the most pressing actions that needs to be taken in the Seto Inland Sea. Therefore, one of the major directions to be taken in the future should be the creative restoration and, possibly, the creative regeneration of a new Seto Inland Sea by the new governance system (Matsuda 2008).

A new concept, "Sato-umi", was proposed by the Research Institute for the Seto Inland Sea. In Japanese, "Sato-umi" means a coastal sea under the harmonization of sustainable, wise use with conservation of an appropriate natural environment and habitat (Yanagi 2007). Compared with a deteriorating coastal environment, "Sato-umi" is able to provide a higher biological diversity as habitat and higher biological production as fishing grounds. Such characteristics are also suitable for demonstrating the multi-functional roles of fisheries and ecosystem services.

Development of a new holistic approach for sustainable biological production and control of eutrophic levels is a prerequisite for establishing functionally efficient "Sato-umi" in each local coastal area. It is recommended that the promotion of integrated environmental management aimed at environmental remediation and restoration of a wide variety of habitats be undertaken in the near future under the new governance system that includes the central government. local government, and a wide variety of stakeholders such as fishermen, navigators, tourist companies, environmental NPO, and so on. This should involve the international exchange of information, ideas, and methodologies.

However, with respect to the future direction of habitat conservation and resource management, top priority should be given to the original objective. In the case of seaweed and seagrass bed restoration, this includes the high performance restoration of seaweed and seagrass beds. However, other viewpoints are also important. Future methods of habitat restoration and resource management should be examined from the viewpoint of low environmental impact, high recycling of material used, low cost with high cost performance, energy saving technology, and applicability of adaptive management. Continuous monitoring after restoration activities is also very important in evaluating the effectiveness of the restoration methods that have been used.

Important and practical future directions are itemized below:

Active conservation of a new environment in the Seto Inland Sea

- Preferable habitat environments and recreational spaces to recover seaweed and seagrass beds, tidal flats, and other shallow water areas must be restored.
- The strict control of reclamation and excavation must be done.
- Promoting fisheries from the viewpoint of the multi-functional role of fisheries to extract desirable ecosystem service.

Regeneration of forests, rivers, and the sea, with effective participation and partnerships among the various stakeholders

- Preferable water and material cycling, recognizing the interactions between forests, river basins, and coastal seas (watershed scale: ecosystem approach) must be established.

Establishment of mitigation systems

- Minimizing waste dumping in the area
- The wise and efficient use of vacant land along the coast
- Securing new environments in areas that are disappearing but of historical significance to recover their ecosystem service

Integrated coastal management (ICM)

- Wide-ranging cooperation among the national government, local governments, local citizens, companies, and other entities. A unified authority to be organized with all rights on management of the Seto Inland Sea

In conclusion, the development of a new holistic approach for sustainable biological production and control of eutrophic levels, or a kind of new creative restoration to recover the many kinds of ecosystem services of the Seto Inland Sea, is a priority. Promotion of integrated environmental management, including watershed management, should be adopted from the viewpoint of interrelated water and material cycling in the river basin, forest, and coastal seas. The concept of "Sato-umi", originating from the traditional ideas of the local people for the wise and sustainable use of coastal areas, can support this new creative restoration of the environment and habitat in the Seto Inland Sea.

Box 3.1: Rehabilitation of Eelgrass Beds by Fishermen

The area of eelgrass beds in the Hinase coastal sea area (the central part of the Seto Inland Sea) decreased after the early 1960s, mainly because of an increase in turbidity and use of agricultural chemicals on land. 590 ha of eelgrass beds in 1945 decreased to 12 ha by 1985 (Fig. Box 3.1). The fish catch by set nets of the Hinase Fishermen Union also decreased at the same time. The fishermen thought that the main reason for the decrease in the fish catch may have been the result of a decrease in eelgrass beds. Fishermen in the Hinase Fishermen Union began the rehabilitation of eelgrass

Box 3.1 (continued)

beds by gathering seeds of eelgrass from the remaining eelgrass beds in spring, preserving them until autumn and scattering the best seeds in late autumn from 1986 on. They have continued this activity for more than 30 years and the area of eelgrass beds expanded to about 80 ha by 2005 and the fish catch by set net also recovered (Fig. Box 3.2). Some local citizens have joined in this activity since 2011 because they believe that the preservation of the marine environment by expanding the area of eelgrass beds is very important for a healthy marine environment in the Seto Inland Sea.

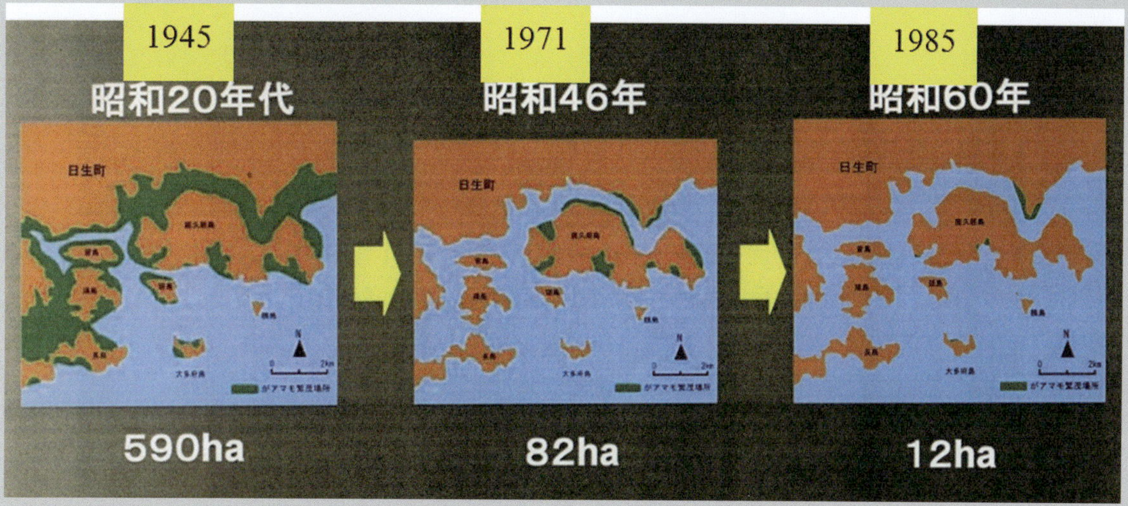

Fig. Box 3.1 Year-to-year variation in the area of eelgrass beds in the Hinase coastal sea

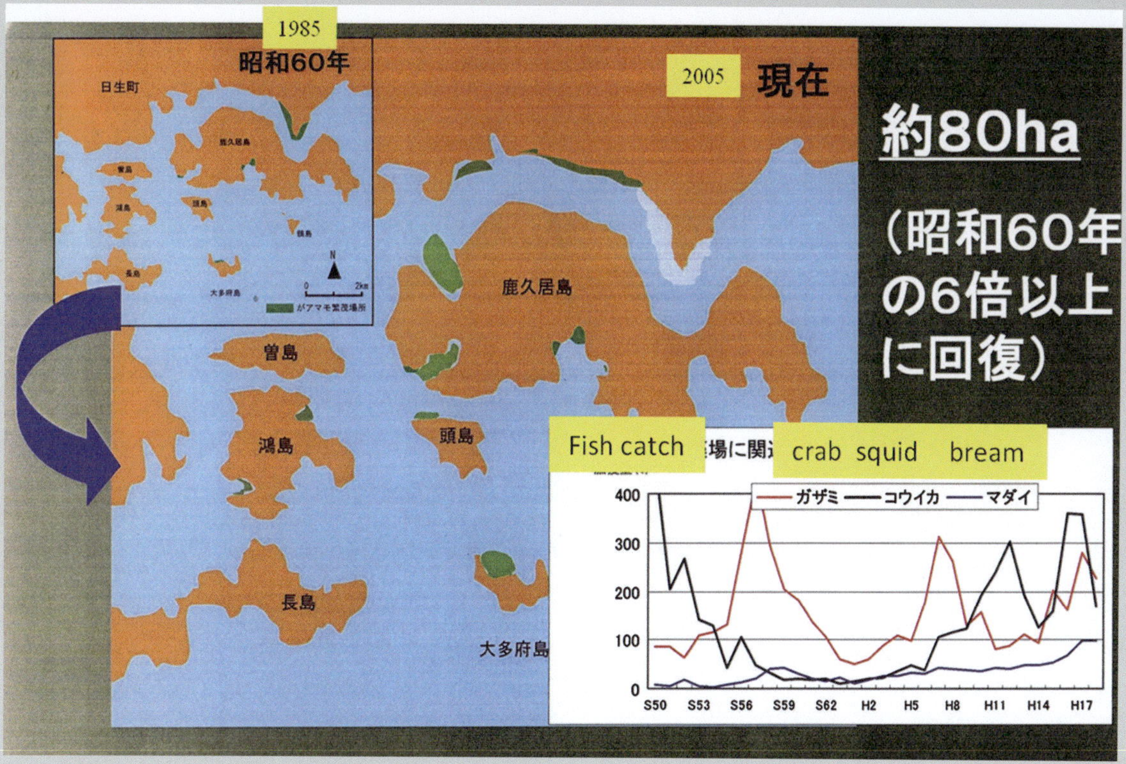

Fig. Box 3.2 Restored eelgrass beds in 2005 and year-to-year variation in fish catch by set nets

Box 3.2: Preservation of Swimming Crab Resources

Young fishermen engaged in the swimming crab fishing in the Tsunemi Fishermen Union (western part of the Seto Inland Sea) became concerned about the decrease in the crab catch after 2001, although they have released many young crabs each year (Fig. Box 3.3). They decided to perform resources management for the crabs under the suggestion of fishery scientists. They began such management by buying swimming crabs with eggs caught in the early summer and released them into the sea, from May 2004 onwards. They obtained funding of one million Japanese yen (approximately 10,000 US dollars) in 2004 to buy such crabs from the Fisheries Basic Fund, Fukuoka Prefecture. The crabs were cultured in water tanks on land, marked with "Do not catch" on their back using permanent marker pens, and were then released into the sea the next day (Fig. Box 3.4). If these crabs were subsequently caught again, the fishermen released them as soon as they saw the mark on their backs. The price was 500 yen per crab, which was half the market price for swimming crab measuring 16 cm (average size at that time of the year), and it was decided to be the same for all caught crabs irrespective of their size. They asked for cooperation in their attempts at resources management from

Box 3.2 (continued)

local fishermen and buyers in the market; thus, the buyer did not buy crabs marked "Do not catch". The target of 2000 caught crabs was achieved in the middle of July 2004. No marked crabs were seen in the market that year. Such activities are still carried on. The catch of swimming crab increased from 127 t in 2003 to 155 t in 2004, to 175 t in 2005 and 206 t in 2006 as the result of this project (Fig. Box 3.3).

Fig. Box 3.4 Marked swimming crab with eggs purchased for release

Fig. Box 3.3 Year-to-year variation in the number of swimming crab juveniles released and their catch

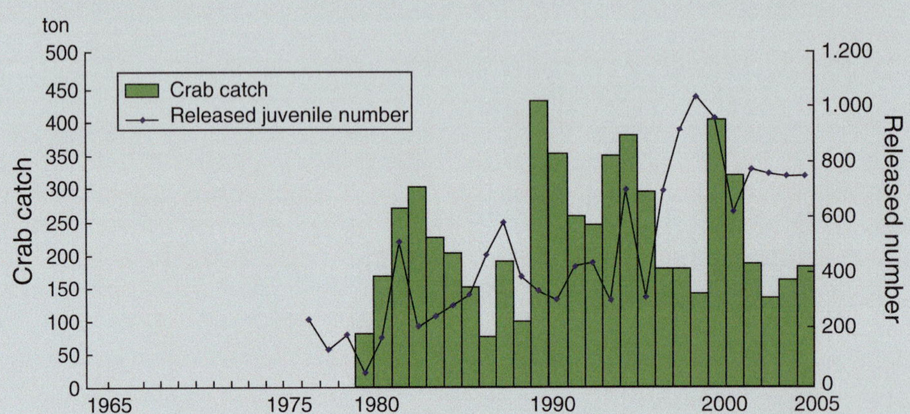

References

Ishii D, Yanagi T (2005) Variation in TP and TN concentrations in the nearshore and offshore areas in the Seto Inland Sea. Oceanogr Jpn 14:35–45 (in Japanese with English abstract and captions)

Ishii D, Yanagi T, Sasakura S (2014) Long-term trends in the occurrence of red tides in the Seto Inland Sea, Japan. Oceanogr Jpn 23:217–236 (in Japanese with English abstract and captions)

Matsuda O (2008) Eutrophication and its causes/consequences: the case of the Seto Inland Sea. In: Mimura N (ed) Asia-Pacific coasts and their management. Springer, New York, pp 78–92

Nishikawa T, Hori Y, Nagai S, Miyahara K, Nakamura Y, Harada K, Tanda M, Manabe T, Tada K (2010) Nutrient and phytoplankton dynamics in harima-Nada, eastern Seto Inland Sea, Japan during a 35 year period from 1973 to 2007. Estuar Coasts 33:417–427

Okaichi T (ed) (2004) Red tides. Terra Scientific Publishing Company, Tokyo, 439 p

Okaichi T, Yanagi T (eds) (1997) Sustainable development in the Seto Inland Sea – from the viewpoint of fisheries. Terra Scientific Publishing Company, Tokyo, 320 p

Tada K, Fujiwara M, Honjo T (2010) Water quality and Nori (Porphyra) culture in the Seto Inland Sea. Bunseki Kagaku 59:945–955 (in Japanese with English abstract and captions)

Tanda M, Harada K (2011) Examples of measures to oligotrophy and a future problem. J Jpn Soc Water Environ 34:54–58 (in Japanese)

Yamamoto T (2003) The Seto Inland Sea – eutrophic or oligotrophic? Mar Pollut Bull 47:37–42

Yanagi T (1997) Budgets of fresh water, nitrogen and phosphorus in the Seto Inland Sea. Ocenogr Jpn 6:157–161 (in Japanese with English abstract and captions)

Yanagi T (2007) Sato-Umi: a new concept for coastal sea management. Terra Scientific Publishing Company, Tokyo, 86 p

Yanagi T (ed) (2008) Sea bottom environment of the Seto Inland Sea. Koseish-Koseikaku, Tokyo, 130 p. (in Japanese)

Yanagi T (2012) Japanese commons in the coastal seas; how the satoumi concept harmonizes human activity in coastal seas with high productivity and diversity. Springer, Tokyo, 113 p

Yanagi T, Tanaka T (2013) Origins of phosphorus and nitrogen in the Seto Inland Sea. Rep Res Inst Appl Mech Kyushu Univ 144:13–18

Disappearance of Hypoxia in Dokai Bay

4

Machiko Yamada

Abstract

Dokai Bay, in northern Kyushu, Japan, though small in size is an important port for the huge heavy and chemical industrial facilities that surround the bay. In the past, it was also a discharge area for heavily polluted industrial waters. Long-lasting hypoxia events were observed over large parts of the bay in the 1990s, but loads of total phosphorus (TP) and total nitrogen (TN) in wastewater have been controlled since 1980 and 1996, respectively, by government regulations. In consequence, TP and TN concentrations derived from nutrients such as phosphate-phosphorus and ammonium-nitrogen have substantially decreased. In addition, the concentration of acid-volatile sulfides in sediments has decreased drastically.

Because of decreased eutrophication, the severity and duration of the hypoxia events have decreased, and, although the red tide formation frequency has not changed, the diversity of red tide organisms has increased. Macroalgae have expanded their distribution to inner and deeper parts of the bay as the light penetration has increased. Sessile animals as well as fish, shrimp, and crabs are significantly more abundant on the bottom of the inner bay because of the increased dissolved oxygen content, especially in summer.

A bioremediation study of the filter-feeding mussel *Mytilus galloprovincialis* at a demonstration facility in the bay revealed that water purification by this mussel reduced the density of red tide organisms. In addition, the facility serves as a gathering place for fishes and as an environmental education site. Moreover, mussels collected from the facility are mixed with wood chips and composted for fertilizer. A seaweed bed bioremediation facility also enriches the ecosystems in the bay. With the disappearance of hypoxic areas, water quality has greatly improved, and it is hoped that restoration of their habitats will cause species that previously lived in the bay to return in abundance. The optimum eutrophic level for the bay still needs to be determined cooperatively by government agencies, stakeholders, and specialists, taking into account the richness of its aquatic resources.

Keywords

Eutrophication • Dokai Bay • Hypoxia • Restoration • Ecological improvement

4.1 Introduction

Because ecological function including aquatic resources and biodiversity, is becoming reduced in the global oceans (Reid et al. 2005, Food and Agriculture Organization of the

M. Yamada (✉)
Fukuoka Woman's University, Kasumigaoka 1-1-1, Higashi-ku, Fukuoka 813-8529, Japan
e-mail: m-yamada@fwu.ac.jp

T. Yanagi (ed.), *Eutrophication and Oligotrophication in Japanese Estuaries*, Estuaries of the World, DOI 10.1007/978-94-017-9915-7_4

United Nations 2005; Heike et al. 2008), the protection of coastal areas from anthropogenic impacts has increased in importance and urgency. Along with overfishing, harmful algal blooms (HABs), toxic contaminants, and physical alteration of coastal habitats, eutrophication is a stressor of coastal ecosystems (Interagency Working Group on Harmful Algal Blooms, Hypoxia, and Human Health 2010 http://www.whitehouse.gov/sites/default/files/microsites/ostp/hypoxia-report.pdf). In addition, hypoxia associated with eutrophication is affecting an increasing number of ecosystems and thus is a major contributor to the decline of coastal ecosystem health and richness that has been observed in recent decades.

A number of bays and coastal areas in Japan suffer from serious eutrophication, similar to coastal areas in other parts of the world (Central Environment Council of Japan 2010). Though countermeasures to eutrophication, such as reducing nitrogen and phosphorus loads to coastal waters, have been implemented for about 30 years, hypoxia and red tide formation, along with declines in fishery catches, still occur, as described in Chaps. 2 and 3 of this book. Dokai Bay in Kitakyushu City, Japan, used to experience annual hypoxic events, but it is now in a state of recovery. In Asia, Dokai Bay is the second most improved bay with regard to the occurrence of hypoxia events, after Tapong Bay in Taiwan (World Resources Institute 2012; http://www.wri.org/project/eutrophication/map). Moreover, Dokai Bay is the best example of the recovery of a semi-enclosed bay in the temperate zone, because Tapong Bay is a lagoon in the tropical zone.

In this chapter, I describe the environmental countermeasures to eutrophication that have been implemented in Dokai Bay along with the resulting changes in water quality and the responses of various organisms to those changes.

4.1.1 Climate and Meteorology

Dokai Bay is located on the northern coast of Kyushu Island, Japan, along the Japan Sea, between latitude 33 52′ 25″N and 33 56′ 25″N and longitude 130 44′ 28″E and 130 50′ 22″E. The present climate is temperate. Based on the data for 30 years between 1982 and 2011, the average air temperature was 16.2 °C with a minimum of −4 °C in January and a maximum of 35.7 °C in August. Average monthly precipitation was 144 mm, with a maximum monthly value of 300 mm in the rainy season (summer) and a minimum of 82 mm in winter. Sunlight hours were high in May and August (about 200 h per month) and low in January (95 h per month) (Japan Meteorological Agency http://www.data.jma.go.jp/obd/stats/etrn/view/annually_a.php?prec_no=82&b).

4.1.2 History and Present Land Use of Coastal Area

Dokai Bay was formed in the Jomon period, 4–5,000 years ago, as a shallow lagoon. Its original shape is shown in Fig. 4.1. This shallow lagoon persisted for several thousand

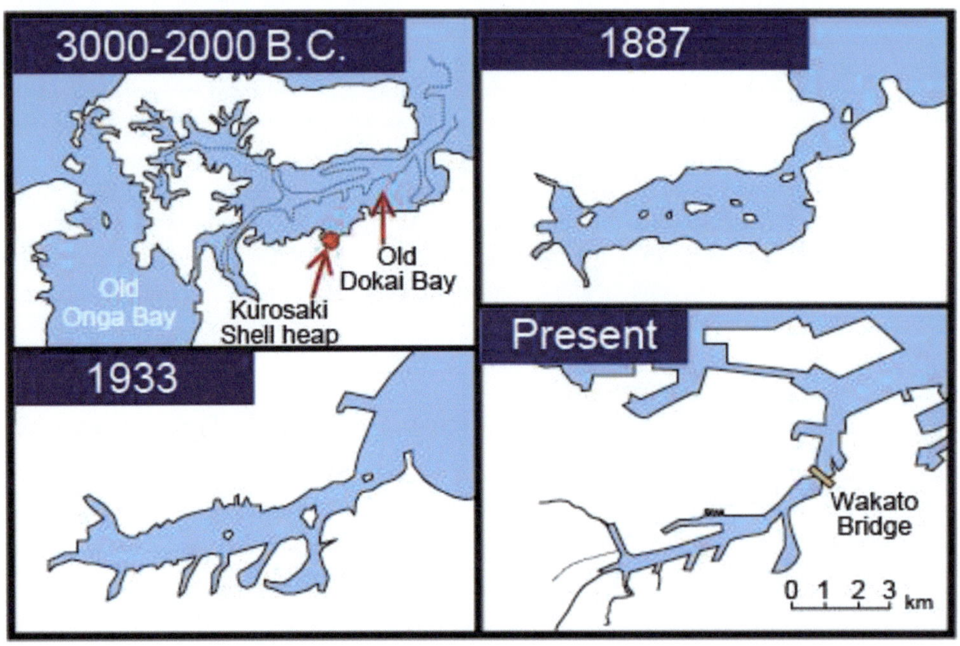

Fig. 4.1 The evolution of Dokai Bay from 3000 to 2000 B.C. to the present

Fig. 4.2 Dokai Bay in 1901 and the modern waterfront

years. After the construction of a government-managed steel plant in 1901, however, its shape was drastically changed by dredging and reclamation, and it became one of the major harbors for heavy and chemical industry in East Asia.

Though the bay's length was extended by the reclamation, the surface area of the bay, about 10 km^2, has not changed. Eventually the bay acquired its present-day length (13 km) and depth (shipping channel depth, 10 m). The width of the bay is 1.2 km at its mouth and 0.3 km in the inner bay. The shoreline, which is 48 km long, consists mostly of artificially constructed sea wall; only 0.5 % is natural shoreline (Fig. 4.2). The shoreline of Dokai Bay is almost entirely lined by industrial development (Fig. 4.3). Five small rivers and a sewage effluent canal flow into the inner bay. The total freshwater inflow from rivers is $20–30 \times 10^4$ m^3 d^{-1} (Higashi et al. 1998).

4.1.3 Fisheries and Heavy Industrial Pollution Before the Onset of Severe Eutrophication

A shell mound (Fig. 4.1, ca. 3000–2000 BC) dating to the Jomon period was discovered along the paleo-shoreline of Dokai Bay, Kitakyushu, Japan (the Kurosaki shell heap). It contained the remains of oysters and fish (porgies and puffers). In the past, Dokai Bay was called "a treasury keeping great Kuruma-ebi prawns" (Yamada 2000). Indeed, since ancient times, Dokai Bay harbored a rich fishery that provided products to the people living in the area. Fishery products including dried sea cucumbers and live eels were shipped to China during the Edo Era and to Kansai during the Meiji Era.

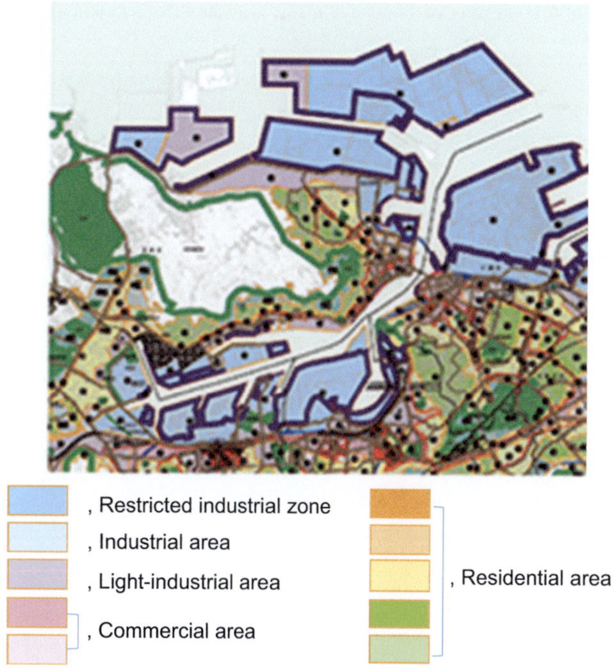

Fig. 4.3 Map of land use in the Dokai Bay area in 2011 (Data for 2012 from the Buildings and City Planning Bureau of Kitakyushu City)

This rich environment, however, changed dramatically after 1901 when the government-managed Yawata Steel Company (currently Nippon Steel & Sumitomo Metal Co., LTD.) was constructed along the shoreline of Dokai Bay (Fig. 4.2). This led to the development of the Kitakyushu heavy chemical industrial zone surrounding Dokai Bay and the subsequent increase in the flow of untreated industrial wastewater into the bay. About 30 years after the construction of the first factory, the total fishery catch decreased by half within 4 years, and in 1933, fishermen around Dokai Bay petitioned the Fukuoka Prefectural Fisheries Experiment Station to investigate the cause of the fishery damage. Bioassays using test fish confirmed the serious acute toxicities of industrial wastewater flowing into the bay (Table 4.1) (Fukuoka Prefectural Fisheries Experiment Station, 1933). Moreover, careful observations by the experiment station confirmed the presence of an azoic zone in the bay.

Since the initial assessment in 1933, industrial pollution by organic compounds and hazardous chemicals expanded throughout the bay with the expansion of the heavy chemical industry. As a result of the serious deterioration of water and sediment quality, fishermen caught nothing after about 1942, although small numbers of fish returned to the bay after factories were damaged during the Pacific War in 1945. Fishery rights in the inner bay were gradually rescinded over time, and by 1951 the fisheries had vanished (Fig. 4.4). Dokai Bay became known as a "dead sea" because it was unable to support living organisms. Such heavy and

Table 4.1 Results of bioassays using seawater from Dokai Bay in 1933

Sampling station	Average time for test fish[a] to die (min)
Near the point of discharge from a glass factory	12
Near the point of discharge from a food factory	34
Near the point of discharge from a sulfuric acid factory	46
The inside of the dock of Factory A	57
The outside of the dock of Factory A	63
Old Wakamatsu Port	75
Near the Tobata fish market	83
Near Wakamatsu light house (unpolluted seawater)	>180

[a]Red sea bream and eel juveniles

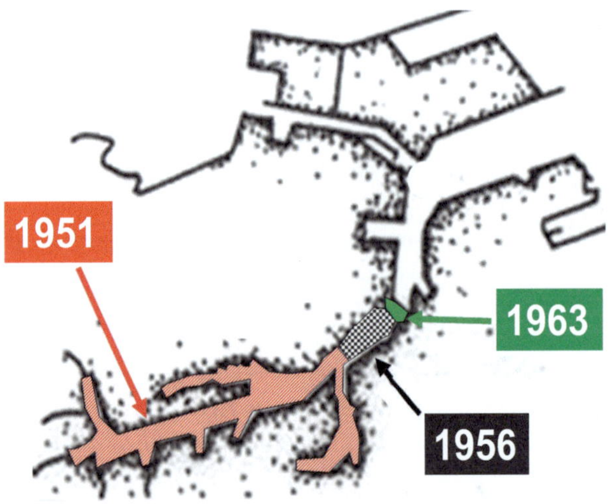

Fig. 4.4 Rescissions of fishing rights from 1951 to 1963 in response to damage from industrial pollution in Dokai Bay

hazardous contamination of the seawater made it easy to strip fouling organisms such as oysters and barnacles from the bottom of shipping vessels entering Dokai Bay.

It was only in 1966 that the national government started to measure water quality of seawater and industrial wastewater in Dokai Bay. The worst observed water quality values from 1966 to 1969 are shown in Table 4.2. The table shows the worst values, not average values, because worse conditions irreparably damaged the bay's organisms and also because water quality was not measured very frequently at that time. The dissolved oxygen (DO) concentration in the seawater measured zero, and both seawater and industrial wastewater were contaminated by many hazardous chemicals, including cyanide and phenol. They were also contaminated and highly turbid with organic or inorganic substances as shown by the high-chemical oxygen demand (COD) and levels of suspended solids (SS).

Table 4.2 Worst values for water quality parameters for seawater in Dokai Bay and industrial wastewater discharged into the bay from 1966 to 1969

Parameter	Units	Worst value	
		Seawater	Wastewater
Transparency	cm		1.0
pH		6.3	3.0/10.1
Dissolved oxygen	$mg\ L^{-1}$	0.0	
Chemical oxygen demand	$mg\ L^{-1}$	74.6	400
Suspended solids	$mg\ L^{-1}$	1,080	2,370
Oils[a]	$mg\ L^{-1}$	5.4	5.5
Phenol	$mg\ L^{-1}$	0.34	45.0
Total cyanide	$mg\ L^{-1}$	0.64	25.0

[a]Normal hexane extracts

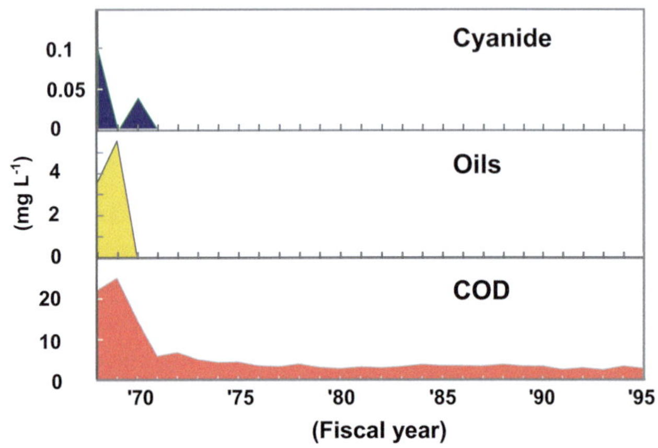

Fig. 4.5 Rapid improvement of water quality at monitoring station D6 (see Fig. 4.13) after the treatment of industrial waste beginning in 1970. Cyanide, Total cyanide. Oils, Normal hexane extracts. COD, chemical oxygen demand

Dokai Bay was finally designated as a water area of concern by the national government, and in 1970, measures were established for the treatment of industrial wastewater before discharge into the bay. As a result, the water quality rapidly improved (Fig. 4.5), and by 1973, all environmental standards of water quality were achieved, contrary to expectations. The major reason for this rapid improvement was that the pollution of Dokai Bay was caused by industrial wastewater discharge from only 20 big companies (Fig. 4.6) that were very amenable to government regulation.

Bottom sediments were also severely polluted with hazardous industrial contaminants including cyanide, mercury, and tar (Table 4.3). The concentrations of three items—cadmium, cyanide, and arsenic—in Dokai Bay were higher than in any other Japanese bay at that time. There was a need to remove the large volume of harmful sediments to prevent not only the risk of accumulation and hazardous effects in organisms but also the navigation hazard caused by thick banks of sediments.

Polluted sediments were dredged from Dokai Bay in 1974 and 1975 by the Kitakyushu Port Management Association (the former Seaport and Airport Bureau of Kitakyushu City) as shown in Fig. 4.7. There was great difficulty in deciding where to dredge because each contaminant had accumulated in different parts of the bay. A total of 350,000 m^3 of sediment containing mercury at 30 mg kg^{-1} dry or higher was eventually dredged from nine mercury-contaminated areas referring to mercury contamination for Minamata disease. Dredged sediments were dumped into a basin reconfigured as a disposal site, which was located in the middle part of Dokai Bay (Fig. 4.8). The dredging cost of six million US dollars at the time was shared by private and public sectors, including the national, prefectural and city governments. Private industry covered about 71 % of the total cost based on the "polluter pays" principle. The sediment quality in the bay improved dramatically, as shown in Fig. 4.8 for total mercury.

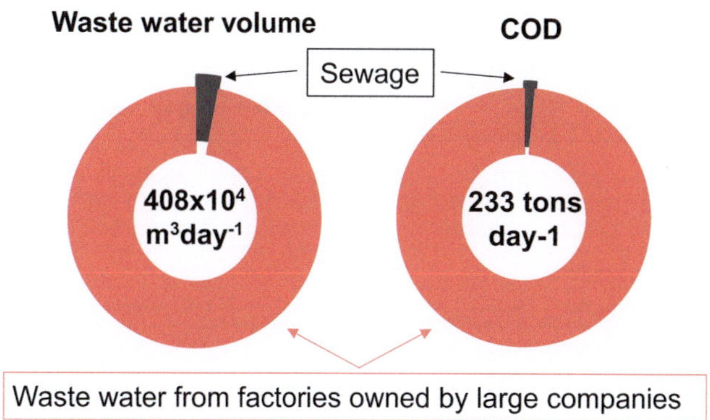

Fig. 4.6 Total wastewater volumes and chemical oxygen demand (COD) loads discharged into Dokai Bay in 1969

Table 4.3 Worst values for pollutants in the sediments of Dokai Bay from 1970 to 1971

Pollutant	Worst value (mg kg^{-1} dry)
Total cyanide	327
Cadmium	603
Arsenic	670
Total mercury	551
Lead	1,870
Total chromium	2,730
Tar	107

The success of the countermeasures for industrial pollution improved the biotic and environmental conditions in the bay. The thriving Kuruma-ebi prawn *Marsupenaeus japonicus* fishery was restarted as a licensed fishery in 1983, reflecting the recovery of the bay from industrial pollution. The pollution and recovery processes were reported in detail by Yanagi et al. (1999) and Yamada (2000). The citizenry, however, was skeptical of the improvement in water and sediment quality because the industrial pollution had been so serious. Therefore, since 1989, there have been ecological investigations into the occurrence of different kinds of producers and consumers, and for water and sediment quality, to demonstrate the recovery. As a result, more than 550 species of living organisms have been confirmed as returned to the bay (Table 4.4). Organisms such as the clam *Ruditapes philippinarum* and phytoplankton form links in food chains as shown in Fig. 4.9. The links between birds and fish or crabs are shown in Box 4.1. Some fish, such as *Konosirus punctatus*, were also shown to reproduce in Dokai Bay, with eggs and juveniles collected as well as adults.

As the above discussion shows, the Dokai Bay ecosystem was confirmed to have recovered over wide areas away from the dead inner part of the bay, but in summer and early autumn, there is a large die-off of many animals in the inner bay, including sessile animals, benthic tidal flat

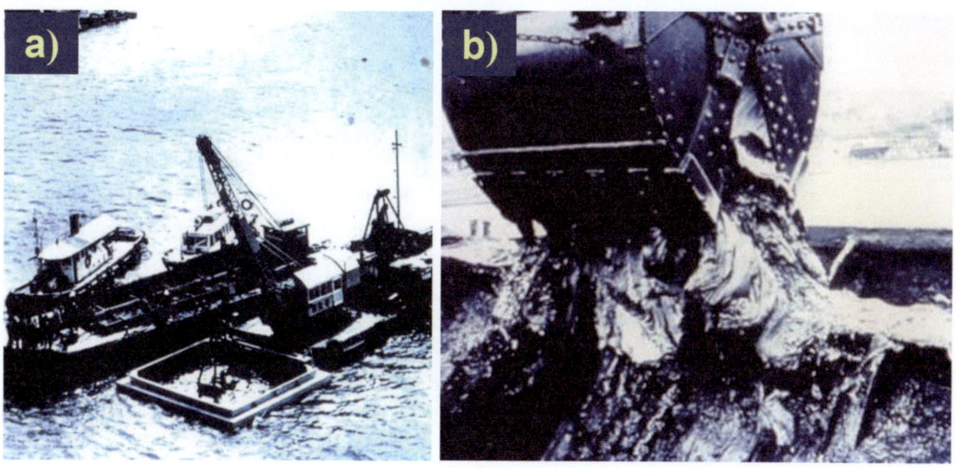

Fig. 4.7 Dredging of polluted sediment in 1974 and 1975 in Dokai Bay. (**a**) Dredging operations, with a box fence preventing the dispersion of polluted sediments, a grab bucket dredger, and a transfer barge. (**b**) Dredging of heavily polluted sediment

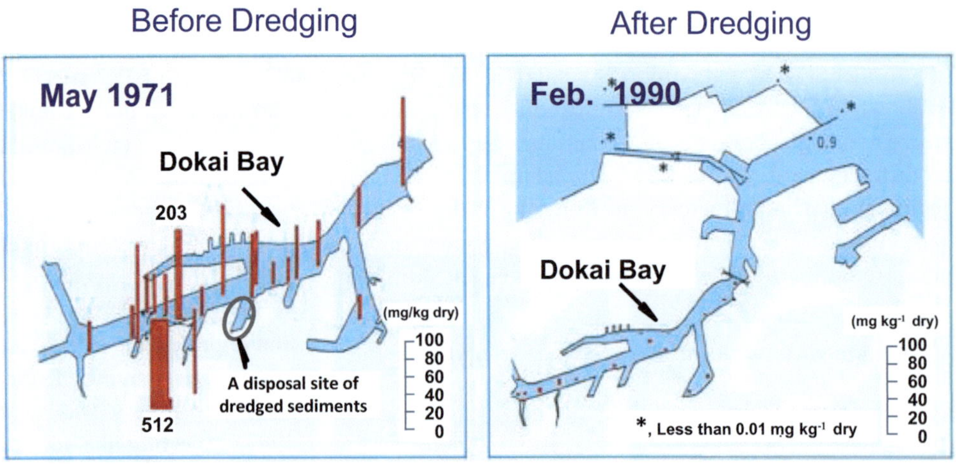

Fig. 4.8 Decrease in total mercury in sediments as a result of dredging

Table 4.4 Organisms confirmed as returned to Dokai Bay from 1989 to1993

Producers		Consumers	
Category	Number of species	Category	Number of species
Phytoplankton	157	Zooplankton	53
Macroalgae	50	Benthic animals	71
		Sessile animals	74
		Fish, shrimp, and crabs	129
		Birds	46

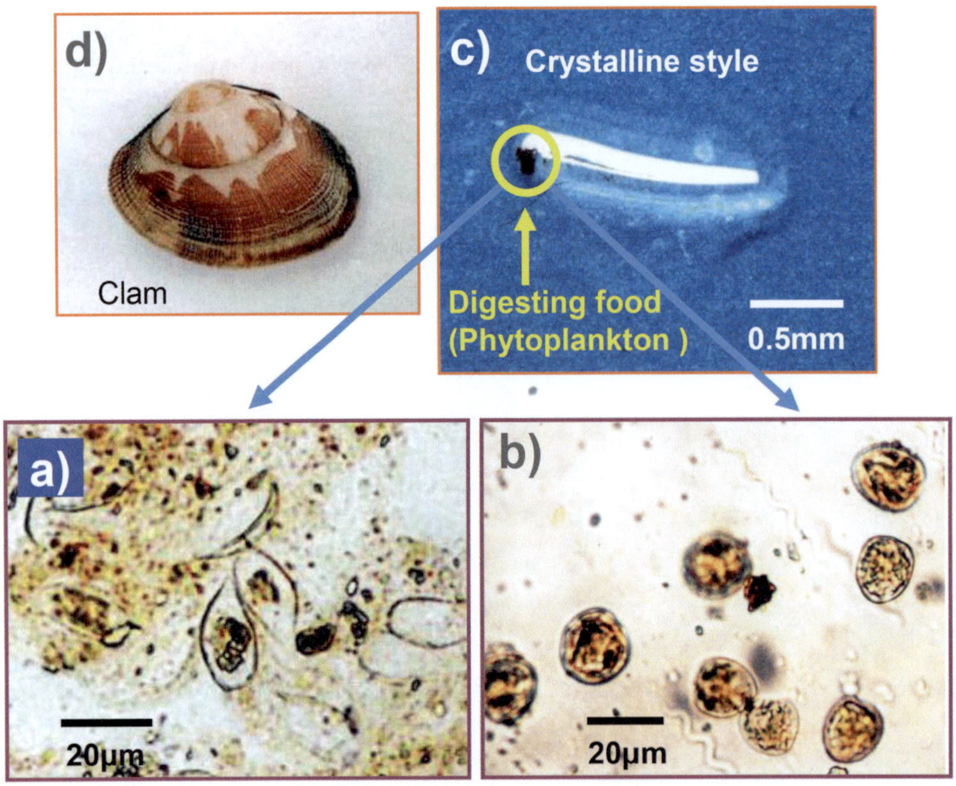

Fig. 4.9 Restoration of a food chain to Dokai Bay in 1973. The dinoflagellates *Prorocentrum triestinum* (**a**) and *P. minimum* (**b**) were observed attached to the end of the crystalline style (**c**) in the digestive system of a clam *Ruditapes philippinarum* (**d**), with each phytoplankton species serving as food when they formed red tides in the bay on different days

Box 4.1: Restoration of the Once Dead Dokai Bay

For most of the time since Dokai Bay formed about 5,000 years ago, people living alongside the bay and biota in the bay have coexisted peacefully. However, about 110 years ago, the bay became the home port of heavy and chemical industries, and all aspects of the bay, including its shape and water and sediment qualities and its biota, as well as the coastal landscape, became drastically different. The damage caused by heavy industrial pollution continued for about 50 years, and at the end of that time, the bay was essentially a dead sea (left photograph).

Through treatment of the industrial waters, however, the water quality was improved within 5 years,

Box 4.1 (continued)

and some organisms returned to the bay within 10 years. Subsequently, however, heavy nutrient loads entering the bay led to eutrophication of the bay, resulting in hypoxic waters and red tides in summer. With regulations limiting the total loads of nitrogen and phosphorus and the restoration of the bay's ecosystems, animals have returned to many areas of the bay (right photographs).

Dokai Bay has recovered from serious pollution at a speed without parallel anywhere else in the world. The history of Dokai Bay thus serves as a valuable testimonial for future generations (Yamada et al. 1990, 2000; Yanagi and Yamada 1999).

Fig. Box 4.1 Dokai Bay in 1960's (*left*) and returned animals there now (*right*)

animals (Fig. 4.10), fish, shrimp, and crabs. These extreme mortality events were shown to result from suffocation from hypoxia caused by eutrophication.

4.2 Changes in Environmental Status

A series of physical, chemical, and biological investigations were carried out to confirm the continuing recovery of the bay ecosystem. These studies confirmed that many different kinds of organisms had returned to the bay, but, in addition, nitrogen and phosphorus concentrations reached maximum values, and hypoxic conditions, mainly at the bottom of the

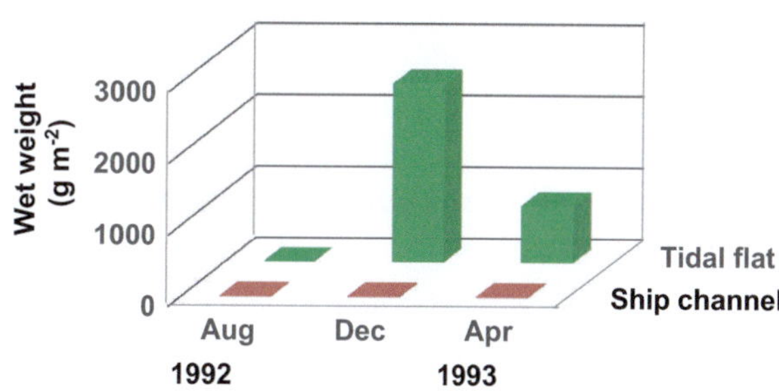

Fig. 4.10 Occurrence of benthic animals in the sediment of a tidal flat and a ship channel in the inner part of Dokai Bay. The biomass of benthic animals in the tidal flat decreased in August to the same level as those in the ship channel in all 3 months sampled

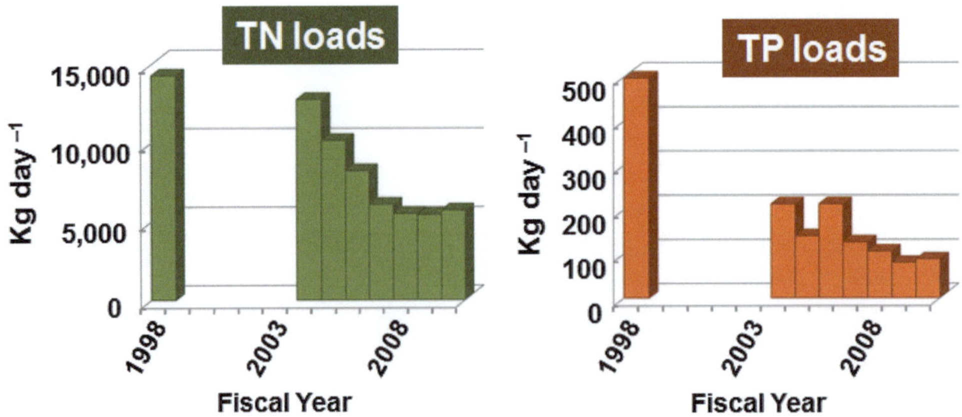

Fig. 4.11 Changes in total nitrogen (TN) and total phosphorus (TP) loading into Dokai Bay

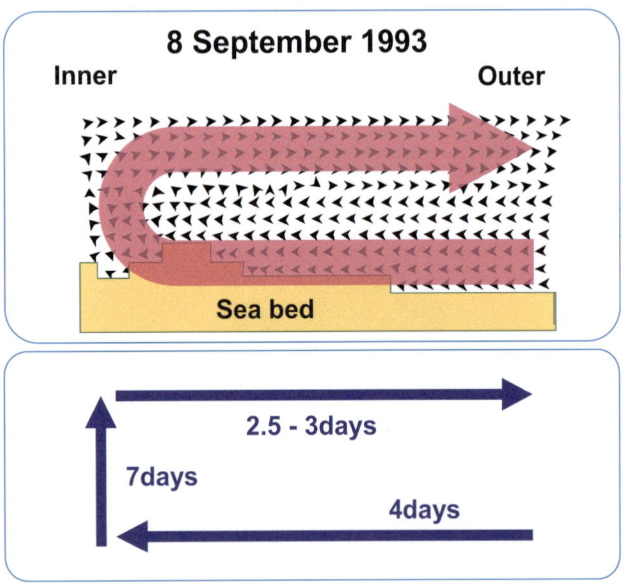

Fig. 4.12 Residual current in Dokai Bay

inner bay, prevented the restoration of fauna, including fish, sessile animals, and macrobenthos (Yamada et al. 1990, Kajiwara and Yamada 1997, Ueda et al. 1994, 2000). Thus, the environmental status of Dokai Bay shifted from polluted by industrial activity to extremely eutrophic.

4.2.1 Governmental Countermeasures Against Eutrophication

Nitrogen and phosphorus loads being discharged into the bay have been reduced at their sources by enforcement of discharge limits established for the Seto Inland Sea District, which includes Dokai Bay. Control of total nitrogen and phosphorus loads was begun in 1980 and 1996, respectively, and data from Kitakyushu City show that both have declined year by year since then (Fig. 4.11) (Record of Environment Bureau of Kitakyushu City). The discharge of domestic effluents accounted for only about one-third and one-fifth of total nitrogen and total phosphorus loads, respectively. Therefore, it was indicated that the industrial effluents were mainly responsible for eutrophic pollution in Dokai Bay.

In the Dokai Bay area, a limit for total nitrogen of 1.0 mg L^{-1} and one for total phosphorus of 0.09 mg L^{-1} were established in 1997. These values are the highest among the four levels designated by the Ministry of the Environment of Japan. To evaluate compliance with these limits, the mean of measured values obtained at two sampling points in Dokai Bay and two sampling points located outside the bay was used. Only one sampling point is usual for determining compliance with environmental standards. In Dokai Bay, however, additional sampling points were used, outside the bay, where nitrogen concentrations were lower than within the bay.

4.2.2 Nitrogen and Phosphorus Residence Times

The maximum tidal range in Dokai Bay is about 1.6 m. The residual current in Dokai Bay (Fig. 4.12) is responsible for most water circulation in the estuary because large volumes of freshwater such as industrial and sewage effluents and riverine waters flow into the inner bay (Fig. 4.13). It takes about 14 days for seawater to make the entire circuit from the mouth of the bay along the bottom to the inner bay, up to the surface layer, and then back out of the bay (Inoue et al. 1998). In 1998, the residence times of freshwater, total nitrogen, and total phosphorus were 7.2, 13.9 and 12.9 days, respectively (Table 4.5) (Yanagi et al. 2001). These residence times are substantially shorter than the corresponding residence times in Tokyo and Osaka bays (Unoki and Kishino 1977; Takao et al. 2004; Yanagi and Takahashi 1988) because Dokai Bay is only about one hundredth the size of the larger bays.

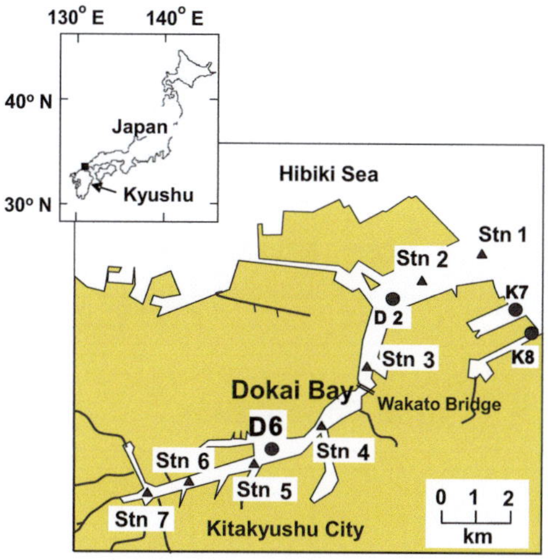

Fig. 4.13 Locations of sampling stations used for water and sediment sampling and for phytoplankton sampling in Dokai Bay

Table 4.5 Residence times of freshwater, total nitrogen (TN), and total phosphorus (TP) in Dokai Bay in 1998

Item	Residence time (day)
Freshwater	7.2
TN	13.9
TP	12.9

4.2.3 Water Quality

Water quality is monitored at several stations in Dokai Bay by Kitakyushu City since 1971 (Fig. 4.13). Time series of long-term changes in six water quality parameters monitored at station D6 are shown in Fig. 4.14. The yearly average and yearly maximum water temperature increased significantly from 1980 to 2010 ($P < 0.001$, $P = 0.005$) at rates of 0.06 °C year^{-1} and 0.07 °C year^{-1}, respectively, but salinity did not change significantly. Transparency increased significantly ($P = 0.001$) at the rate of 0.03 m year^{-1}. Chlorophyll *a*, however, showed no significant trend. Total nitrogen (TN) and total phosphorus (TP) both decreased significantly ($P < 0.001$, $P < 0.001$); the yearly TN average of 9.0 mg L^{-1} in 1992 decreased to 1.0 mg L^{-1} in 2010, and the yearly TP average of 0.24 mg L^{-1} in 1982 decreased to 0.06 mg L^{-1} in 2010. The measured values of both TN and TP at station D6 met environmental standards by 2010 (see Sect. 4.2.1) but at two outside stations. TN began to decline (Fig. 4.14) before total volume controls on nitrogen loads were established in 1996, because industries around the bay began to treat their wastewater in anticipation of the future TN controls.

The distributions of nutrient concentrations at seven stations, from Stn 7 in the inner bay to Stn 1 at its mouth (Fig. 4.13), before (1995–1998) and after (2006–2009) when the environmental standards were enforced are compared in Fig. 4.15 (Hamada et al. 2010, 2012). The concentrations of three nutrients, ammonium-nitrogen (NH_4-N), nitrate plus nitrite nitrogen (NO_3-N + NO_2-N), and phosphate-phosphorus (PO_4-P), decreased from Stn 7 to Stn 1 during both periods. After enforcement of the environmental standards for nitrogen and phosphorus begun, concentrations of both NH_4-N and PO_4-P became much lower in the inner and middle bay, and NO_3-N + NO_2-N were also lower in the middle and outer bay compared with 1995–1998. In contrast, silicate-silicon ($SiOH_4$-Si) concentrations did not differ significantly between the two periods. These results show that the countermeasures against TN and TP loading were effective.

Concentrations of particulate organic carbon (POC), particulate organic nitrogen (PON), and chlorophyll *a* (Chl. *a*) in the inner and middle bay did not differ greatly before and after enforcement of the standards, but they were lower in the outer bay after enforcement (Fig. 4.16) (Hamada et al. 2010, 2012). The concentration of particulate phosphorus (PP) also decreased after enforcement began.

4.2.4 Sediment Quality

Sediment quality was investigated in September of 1995 and 2008 (Ueda, personal communication) (Fig. 4.17). Concentrations of both total organic carbon (TOC) and total organic nitrogen (TON) were the highest at Stn 6 in the inner bay, and they decreased rapidly toward the mouth of the bay. TOC and TON concentrations differed little between the 2 years. The C:N ratio of the sediment did not differ before and after enforcement of the environmental standards, as it was 25.4 in 1995 and 28.0 in 2008. Acid-volatile sulfides (AVS) in the inner bay decreased notably in 2008 compared with 1995, indicating recent improvement of the anaerobic bottom environment.

4.3 Responses

Not only has the eutrophic environment of Dokai Bay been improved, the DO content and the species composition of different taxonomic groups have changed in response to the enforcement of environmental standards.

4.3.1 Reduction and Disappearance of Hypoxia

Examination of temporal changes in dissolved oxygen (DO) in relation to depth during summer 1994 (Fig. 4.18) (Higashi et al. 1998) indicates that hypoxia, defined as DO < 3 mg L^{-1}, was present throughout the summer and severe hypoxia, with DO < 1 mg L^{-1}, occurred in late July. Comparison with the vertical DO distribution in longitudinal

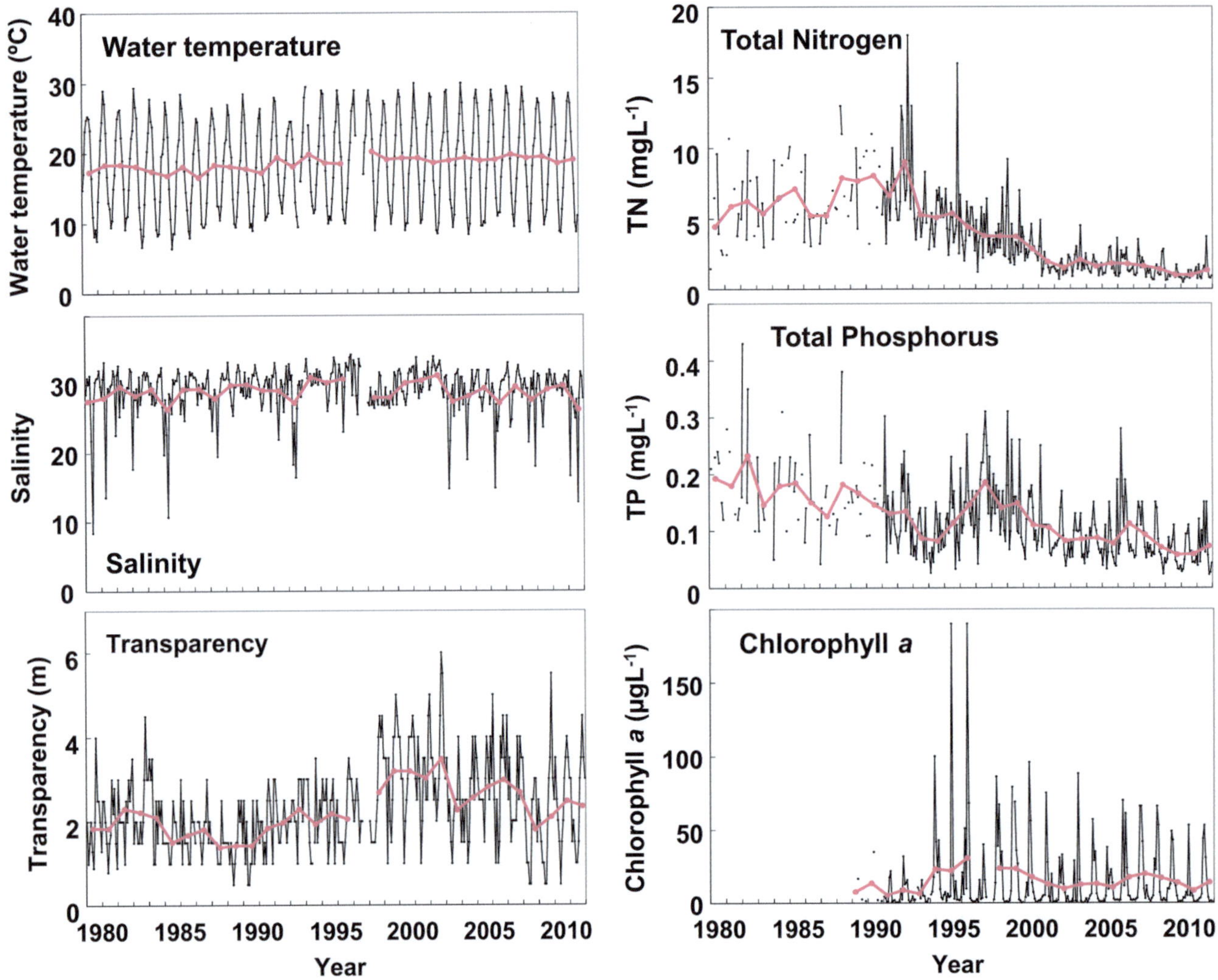

Fig. 4.14 Changes in surface water quality parameters at a governmental monitoring station D6 (Fig. 4.13) from 1980 to 2011. Red lines are connected annual average values

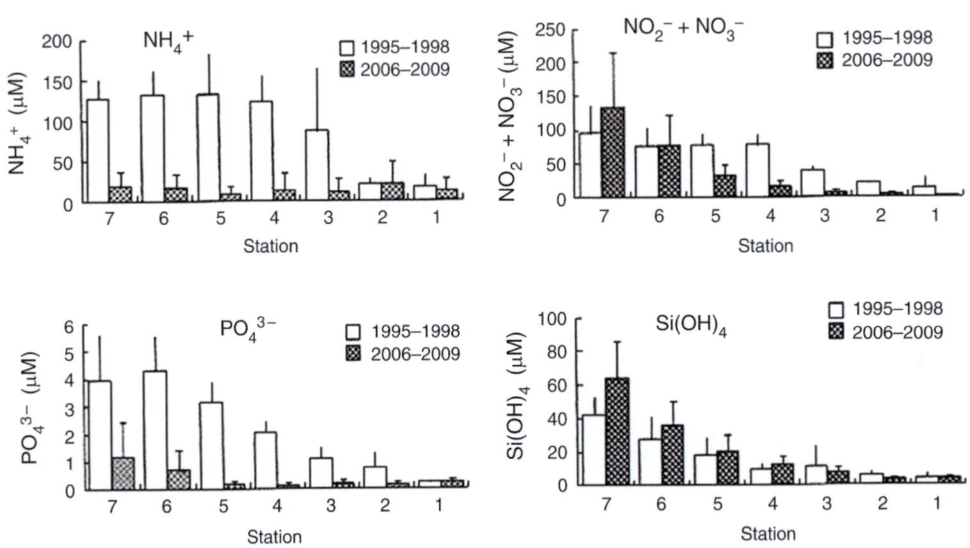

Fig. 4.15 Nutrient concentrations in surface water at seven stations (Stns 1–7; Fig. 4.13) in Dokai Bay in 1995–1998 and in 2006–2009. Error bars indicate the standard deviation (after Hamada et al. 2010, 2012)

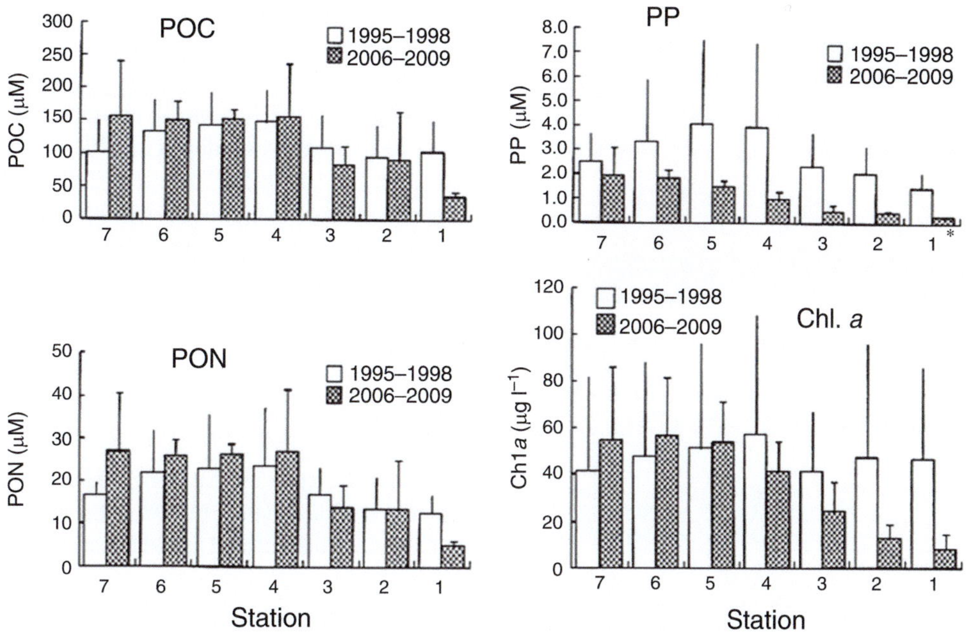

Fig. 4.16 Particulate organic carbon (POC), particulate organic nitrogen (PON), particulate phosphorus (PP), and chlorophyll *a* (Chl. *a*) concentrations in surface water at Stns 1–7 (Fig. 4.13) in Dokai Bay in 1995–1998 and in 2006–2009. Error bars indicate the standard deviation (after Hamada et al. 2012)

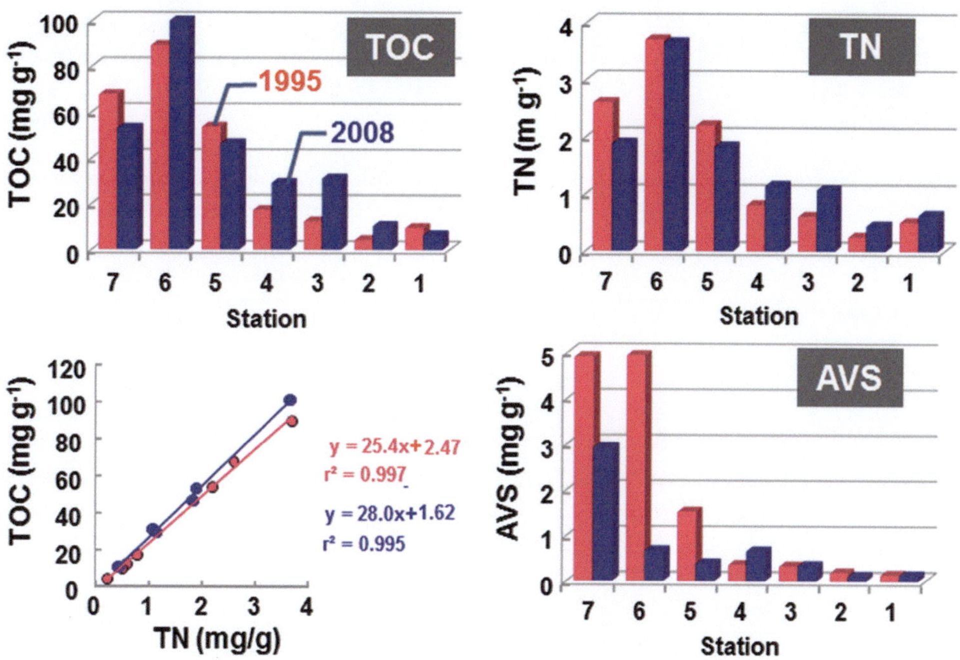

Fig. 4.17 Total organic carbon (TOC), total organic nitrogen (TON), and acid-volatile sulfide (AVS) in surface sediments (0–1 cm) at Stns 1–7 (Fig. 4.13) in Dokai Bay in September 1995 and 2008

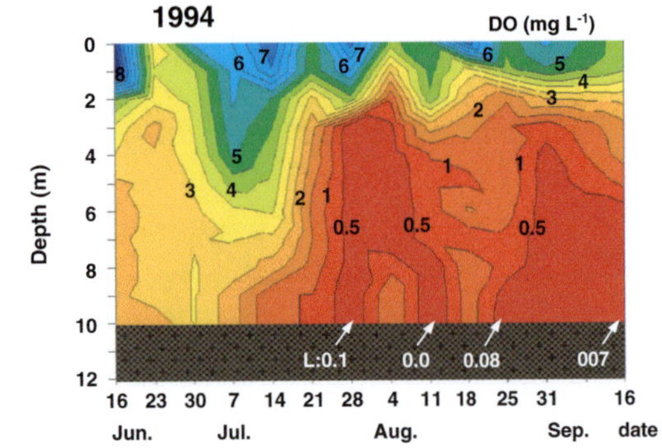

Fig. 4.18 Temporal variation of dissolved oxygen (DO) at Stn 7 in the inner part of Dokai Bay from June to September in 1994

sections during the summers of 1994 (Higashi et al. 1998), 2006, and 2011 (Ueda and Hamada, personal communication) (Fig. 4.19) shows that in 1994, hypoxic conditions were spreading from the bottom of the inner bay to the middle bay. The hypoxia was reduced in 2006, and it disappeared in 2011. In Fig. 4.19, the data for 2006 were collected in September, but no hypoxia was observed just 2 months earlier (July and August in 2006). These results suggest that not only the severity of the hypoxia but also the duration of hypoxic conditions have declined year by year.

4.3.2 Red Tides

In Dokai Bay, red tides characteristically occur for half a year, centered around August. In 1980, the summer was cold and rainfall was abundant, and no red tide occurred (Fig. 4.20). The

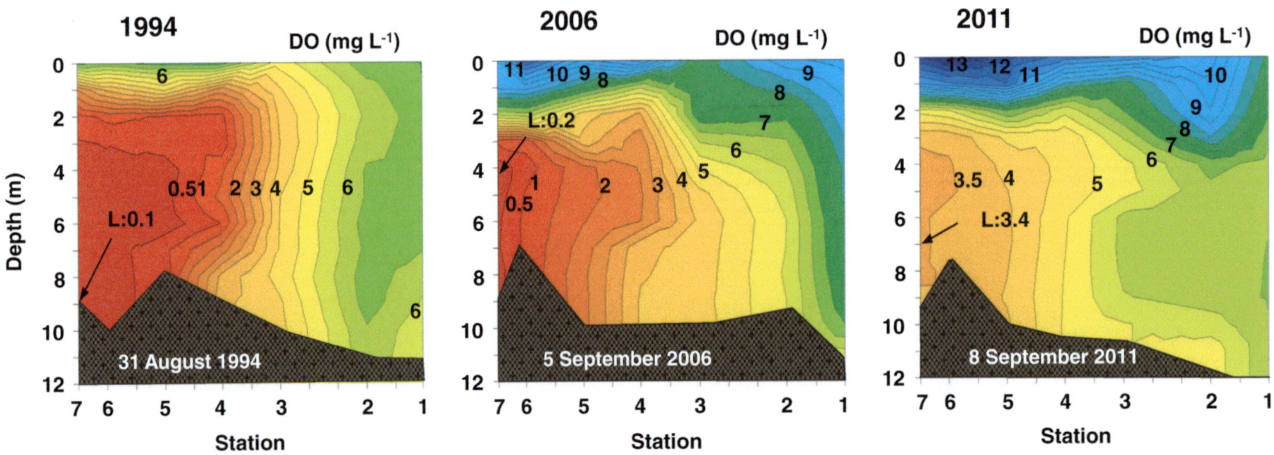

Fig. 4.19 Temporal changes in the longitudinal section of Dokai Bay showing the distribution of dissolved oxygen in summer from 1994 to 2011

Fig. 4.20 Occurrence of red tides in the surface water monitored at station D6 in Dokai Bay from 1980 to 2012. Red dots indicate months when a red tide (Chl. *a* > 10 μg L^{-1}) was observed

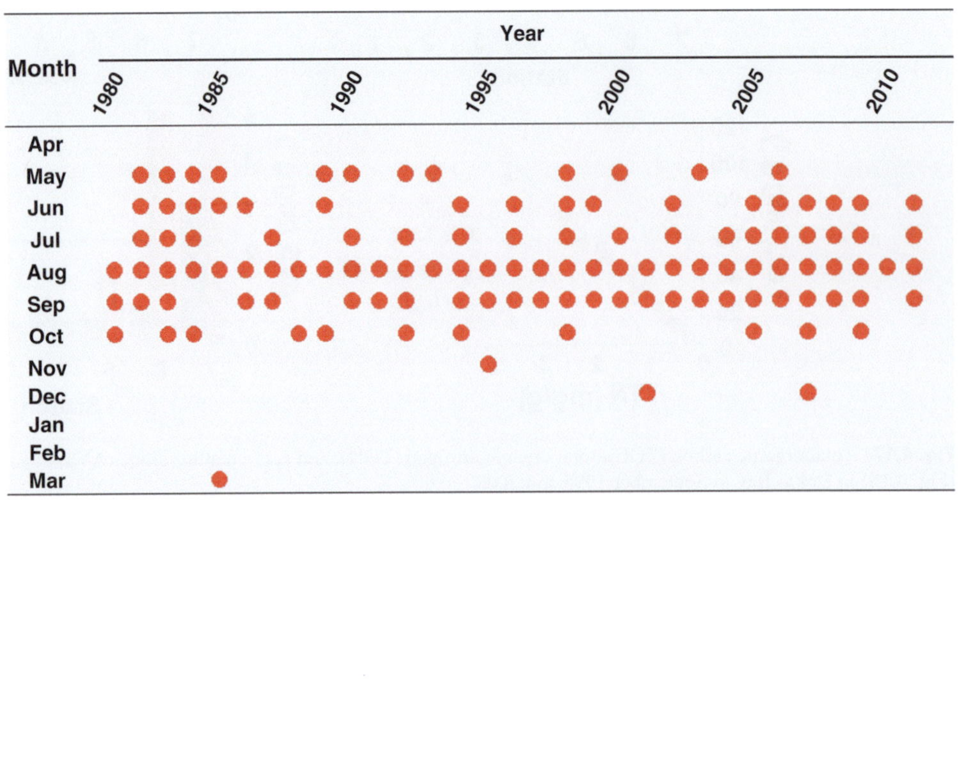

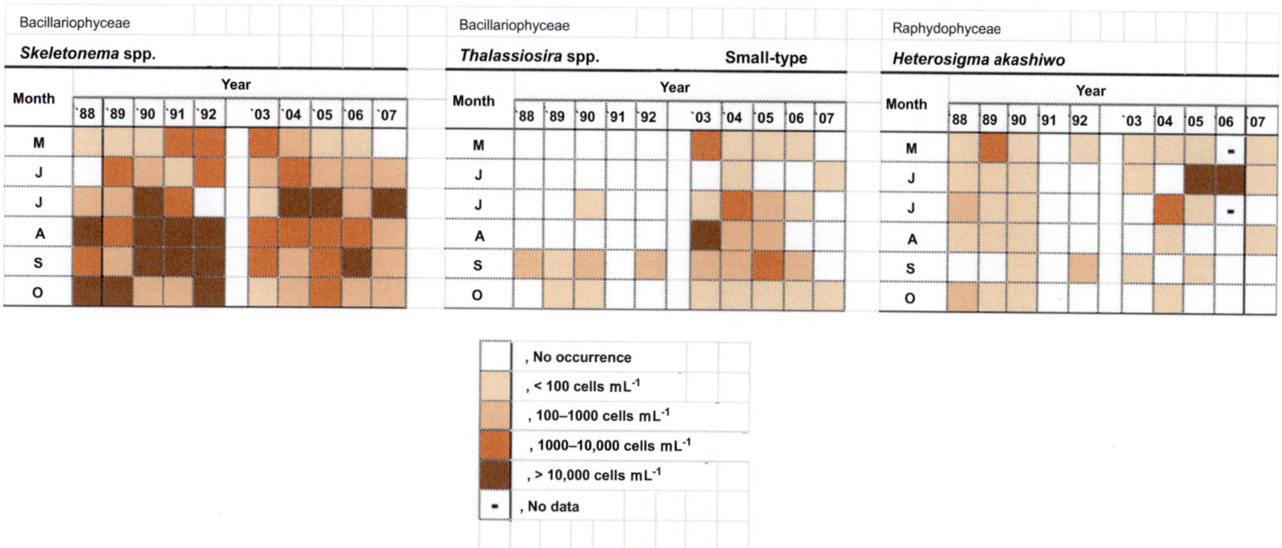

Fig. 4.21 Occurrence of three representative red tide organisms in surface water at monitoring station D6 in Dokai Bay during periods of high (1988–1992) and low (2003–2007) N and P concentrations

Fig. 4.22 Abundance of viable resting-stage cells and cysts of three representative red tide organisms in sediments collected from Dokai Bay in 1997, 1998, and 2008. MPN, most probable number (log [MPN g^{-1}]); AVS, acid-volatile sulfide (mg g^{-1})

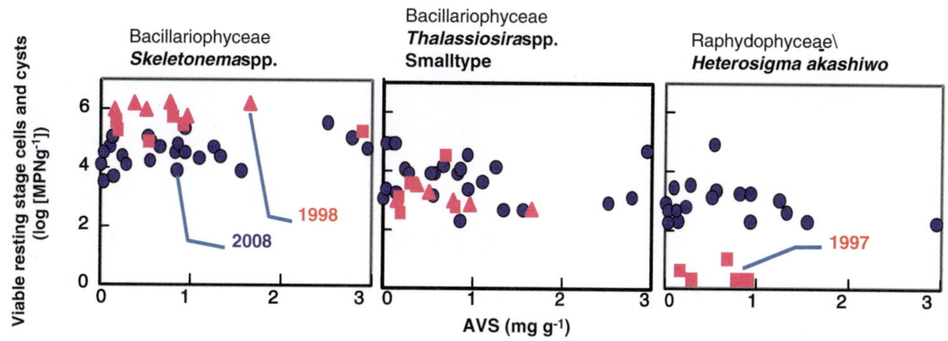

average rainfall during May to October in 1980 was 375 mm $month^{-1}$, which reached about 2 times of the one from 1982 to 2010 (Japan Meteorological Agency). Thus, red tides form in high-temperature periods when the phytoplankton division rate exceeds the rate at which the phytoplankton are carried out of the bay by estuary circulation (Yanagi and Yamada 2000; Tada et al. 2004, 2007).

Red tides in Dokai Bay are primarily due to diatoms, and less often to dinoflagellates and Raphidophyceae; toxic dinoflagellates are especially rare. From 1980 to the present, diatoms of genus *Skeletonema* have been dominant among red tide organisms, but their dominance has decreased with the reduction of TN and TP concentrations. The diatom *Thalassiosira* spp. and *Chaetoceros* spp. and the Raphidophyte *Heterosigma akashiwo* formed red tides this century. As a result, the species diversity of red tide organisms has increased in recent years (Fig. 4.21) (Yamada et al. 2011).

The change in the species composition of red tide organisms is supported by changes in the abundance of physiological resting-stage cells and cysts of different organisms. Fewer viable resting-stage cells *Skeletonema* germinated this century, whereas those of *Thalassiosira* and cysts of *Heterosigma akashiwo* showed increased germination, forming red tides in recent years (Fig. 4.22) (Yamada et al. 2011).

Skeletonema tropicum can be distinguished from other *Skeletonema* species under a light microscope because *S. tropicum* has more than three chloroplasts per cell (Hulburt and Guillard 1968; Sarno et al. 2005) (Fig. 4.23). When eutrophication was severe, *S. tropicum* was one of the most dominant red tide organisms. For example, a widespread, dense red tide composed of *S. tropicum* was observed in August 1994 (Fig. 4.24, after Yamada et al., 2009). Because *S. tropicum* is a tropical species, it is most likely to form a red tide in Dokai Bay in August; this species

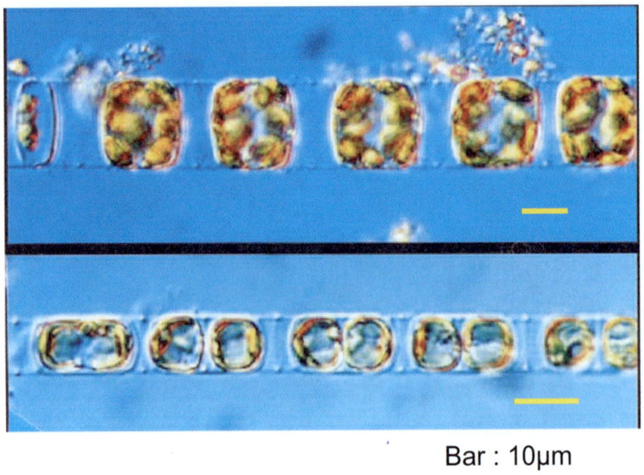

Fig. 4.23 The dominant red tide organisms in Dokai Bay: *Skeletonema tropicum* (upper, diatom) and *Skeletonema costatum* sensu lato (lower, diatom)

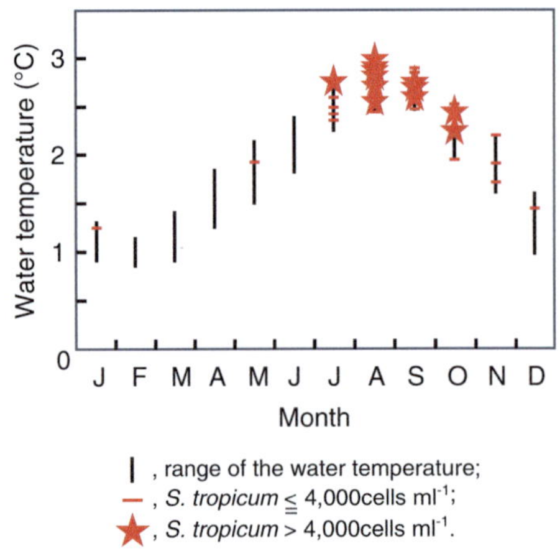

Fig. 4.25 Occurrence of *S. tropicum* in surface water at station D6 in Dokai Bay during 1991 and 2006

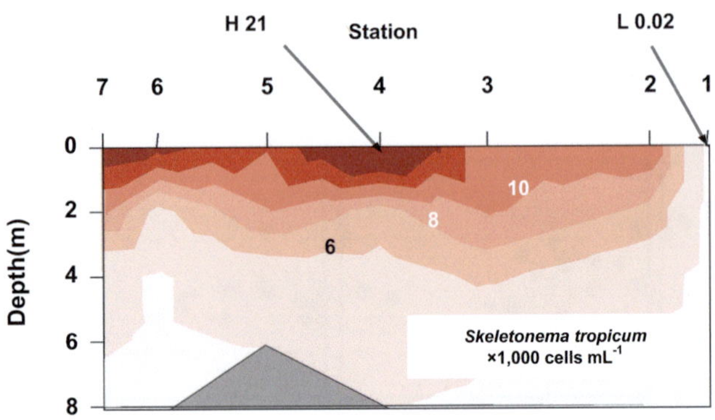

Fig. 4.24 Longitudinal section of Dokai Bay showing the distribution of *S. tropicum* cell density on 31 August 1994. H, highest value; L, lowest value

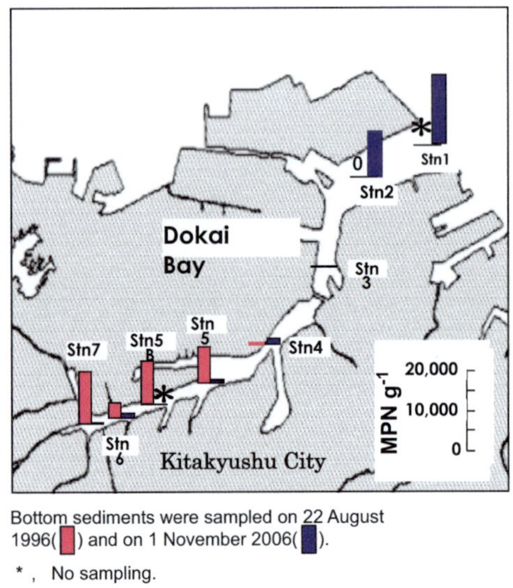

Fig. 4.26 Distribution of viable resting-stage cells of *S. tropicum* in the sediments of Dokai Bay

is seldom observed during the months from December to June (Fig. 4.25). The occurrence period of *S. tropicum* is consistent with the result shown by Hulburt and Guillard (1968) that *S. tropicum* is unable to live at temperatures lower than 13 °C. Culture experiments have shown that *S. tropicum* can germinate from resting-stage cells in the sediments from the entire bay (Fig. 4.26) under appropriate light and temperature conditions (Table 4.6). Overwintering of a tropical species *S. tropicum* by forming physiological resting cells enabled it to establish itself in Dokai Bay located temperate zone (Yamada et al. 2009).

The relationship between water properties and *S. tropicum* cell density shows the ecological features of the preferred habitat of *S. tropicum* (Fig. 4.27). *Skeletonema tropicum* can form a red tide when NH_4-N is as high as 800 μM, although at this concentration some red tide organisms, including the Raphidophyte *Chattonella antiqua*, cannot survive. These data show that *S. tropicum* can adapt successfully to extreme eutrophic conditions such as those that were previously found in Dokai Bay. Suksomjit et al. (2009) reported that the growth of *Skeletonema* spp. in Dokai Bay was accelerated by the presence of ammonium, which indicates that species of this genus are well adapted to eutrophic conditions.

Genetic analyses and observations of fine morphology have shown that the genus *Skeletonema* comprises 11 species (Zingone et al. 2005, Sarno et al. 2005, 2007). As many as seven *Skeletonema* species have been observed in Dokai Bay (Fig. 4.28), which is the highest number of *Skeletonema*

Table 4.6 Temperature and light conditions required for germination of resting-stage cells of *Skeletonema tropicum*

	Light	
Temperature (°C)	24 h Dark	14 h Light[a] and 10 h Dark
10	–	–
11	–	–
23	–	+
26	–	+

[a]60 μmol m^{-2} s^{-1}

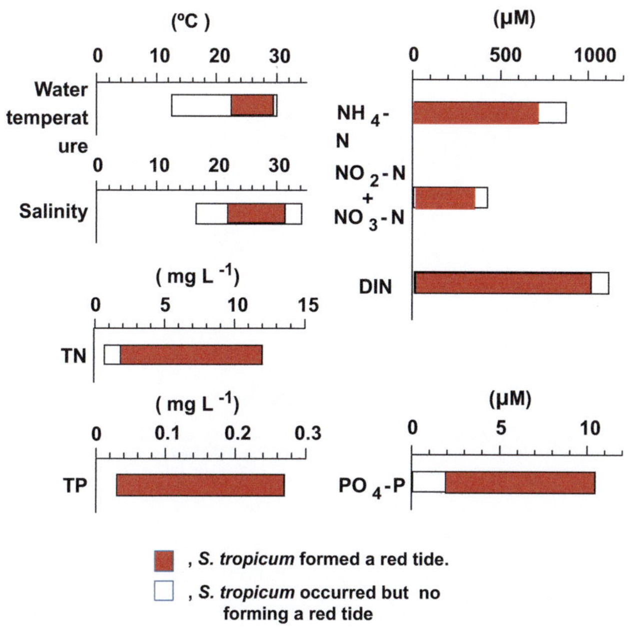

Fig. 4.27 Ranges of water quality parameters for occurrences of *S. tropicum* vegetative cells in Dokai Bay during 1991 and 2006

species that has been observed in coastal areas globally (Yamada et al. 2010).

4.3.3 Macroalgae

Macroalgal compositions in Dokai Bay were compared between May 1991, during the period of severe eutrophication (Yamada et al. 2005), and May 2007, after the eutrophic conditions had improved (Hamada and Ueda, personal communication) (Fig. 4.29). The number of species and the abundance of brown and red algae, including Japanese kelp *Undaria pinnatifida*, a popular edible species, increased and expanded their habitat from the outer to the middle parts after the extremely eutrophic environment was improved. Thus, the reduction of extreme eutrophication enriched macroalgal diversity and biomass.

Transparency, an indicator of the level of eutrophication, was compared between 1991–1995 and 2007–2011, two periods with low Chl. *a*, from December to May, when macroalgae flourish (Table 4.7). Transparency measured using a Secchi disk was higher by 1.1 m during the latter period compared with the former, even though Chl. *a* concentrations were similar in both periods. These results suggest that suspended matter other than phytoplankton decreased with the reduction in the eutrophication level. Moreover, irradiance into the water increased horizontally as well as vertically. The euphotic zone deepens by about 3 m with a transparency increase of 1 m (Hashimoto and Tada 1997), so the euphotic zone in Dokai Bay deepened by about 3 m between the two periods.

Fig. 4.28 Seasonal changes of seven *Skeletonema* species, identified by genetic analyses (LSU rDNA and SSU rDNA) and by their fine morphological structures, in Dokai Bay. Box height indicates the number of strains isolated of each species

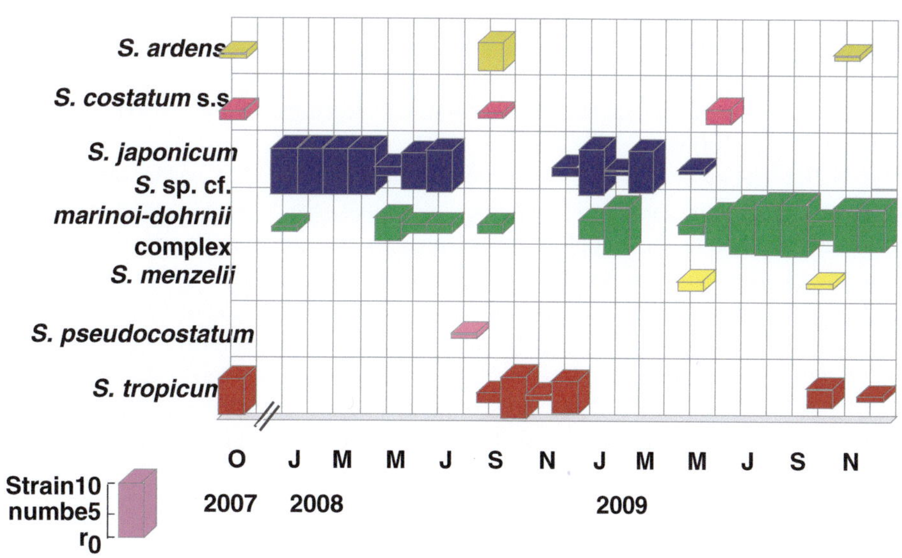

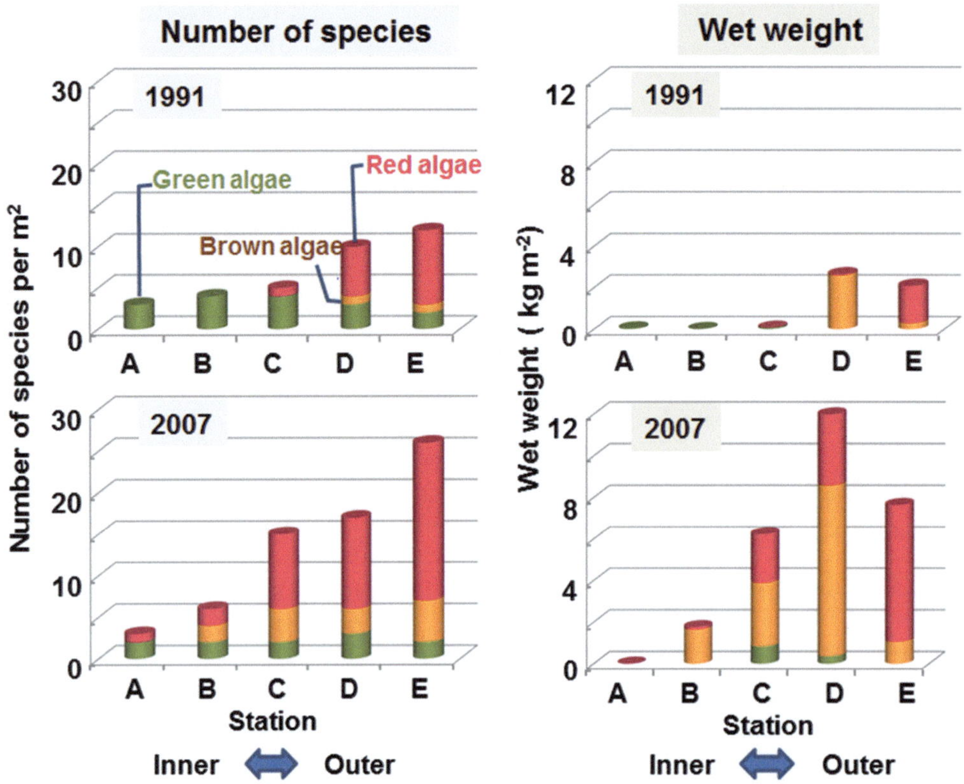

Fig. 4.29 Differences in the occurrence of macroalgae in Dokai Bay among Stations A–E during periods of high (May 1991) and low (May 2007) N and P concentrations

Table 4.7 Transparency from December to May, when Chl. *a* was less than 5 μg L^{-1}, during periods of high (1991–1995) and low (2007–2011) N and P concentrations

Fiscal year	Transparency average (m)	Chl. *a* average (μg L^{-1})	*n*
1991–1995	2.2	2.0	18
2007–2011	3.1	1.9	23

4.3.4 Sessile Animals

As representative sessile animals, mollusks were studied in August 1991 (Kajiwara and Yamada 1997) and 2010 (Fig. 4.30). The number of species of mollusks increased at all sampling stations; gastropod (snail) species became especially abundant on the bottom of the inner bay in 2010. New settlements of gastropods were attributed to there being sufficient DO around bottom quay wall areas of the inner bay (Fig. 4.31)

On the other hand, the total wet weight of mollusks in 2010 was vastly reduced compared with that in 1991. The steep decline was primarily due to decreases in the wet weight of the dominant sessile animals such as the mussel *Mytilus galloprovincialis* and the oyster *Crassostrea gigas* (Fig. 4.32). As Chl. *a* did not change between the two periods, a lack of food cannot explain the decrease. Because the decrease in the wet weight of mussels was associated with an increase in water temperature (Fig. 4.14), the recent rise of water temperature might explain the dramatic reduction of mussel populations (Fig. 4.33).

4.3.5 Fish, Shrimp, and Crabs

The spatial distribution of fish, shrimp, and crabs collected with a small trawling net trawled for 10 min was investigated in August, when hypoxia was likely, in 1989 (Yamada et al. 1990) and 2006 (Ueda and Murata, personal communication) (Fig. 4.34). The numbers of species in 2006 were slightly higher than the numbers in 1989 in the inner bay, whereas the wet weight at each sampling station was much greater in 2006 than in 1989. In the inner bay, the wet weight increased 70-fold. The recovery of DO in recent years likely

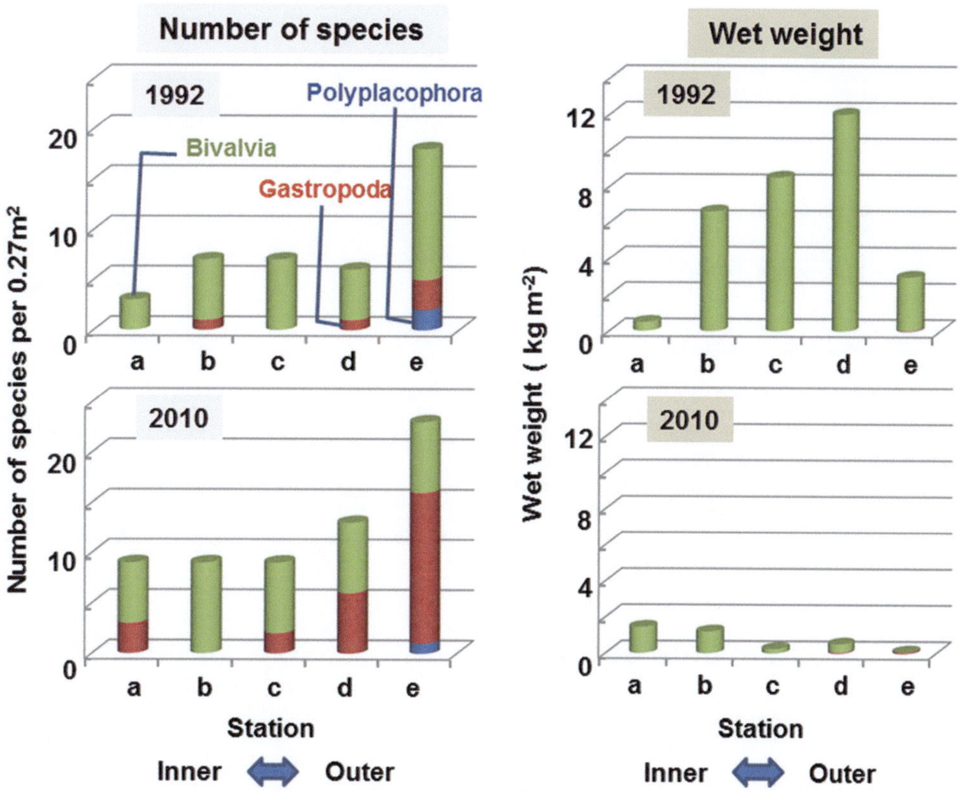

Fig. 4.30 Differences in the occurrence of mollusks in Dokai Bay among Stations a–e during periods of high (August 1992) and low (August 2010) N and P concentrations

Fig. 4.31 Dissolved oxygen concentration conditions in Dokai Bay when sessile animal investigations were carried out in summer of 1992 and 2010

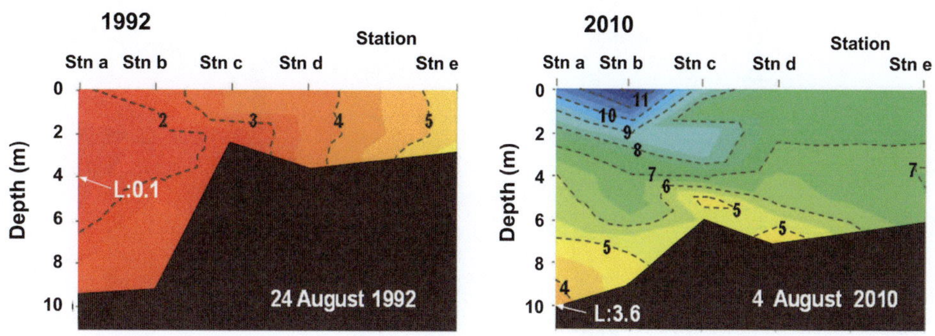

led to an increase in the biomass of animals such as the mantis shrimp, *Oratosquilla oratoria*.

4.4 Future Tasks

In 1998, we proposed an "ecological environmental restoration" method (Yamada et al. 1998) to manage and improve environmental conditions, including TN and TP concentrations, and to enhance biodiversity and biological resources such as fishery species (Fig. 4.35).

4.4.1 Ecological Environmental Restoration

The ecological restoration method consists of five components (Yamada et al. 1998). The objective of the first and second components requires water purification: the first component is the cultivation of filter-feeding bivalves in the warm season (Montani et al. 1998) and the second is the cultivation of macroalgae. The mature bivalves and macroalgae are collected and utilized as resources; the third component is recycling (Fig. 4.36). The fourth component involves sediment purification, as follows: mud-feeding

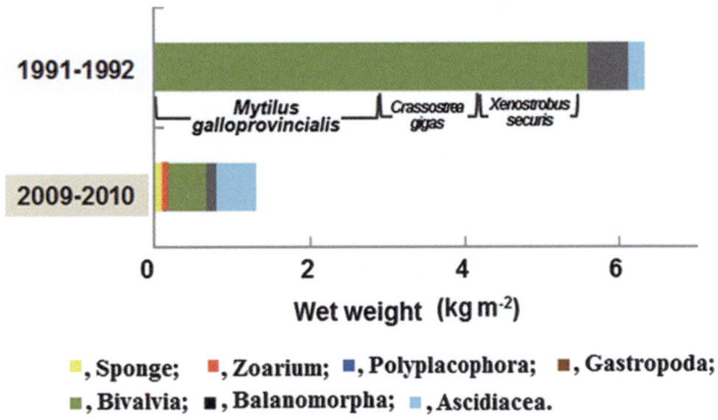

Fig. 4.32 Comparison of average wet weight per sampling point per time of sessile animal taxa in Dokai Bay. The sampling was conducted at three vertical sampling points (supratidal zone, subtidal zone, and near-bottom) at each of five stations in the bay and in each of the four seasons during periods of high (1991–1992) and low (2009–2010) N and P concentrations

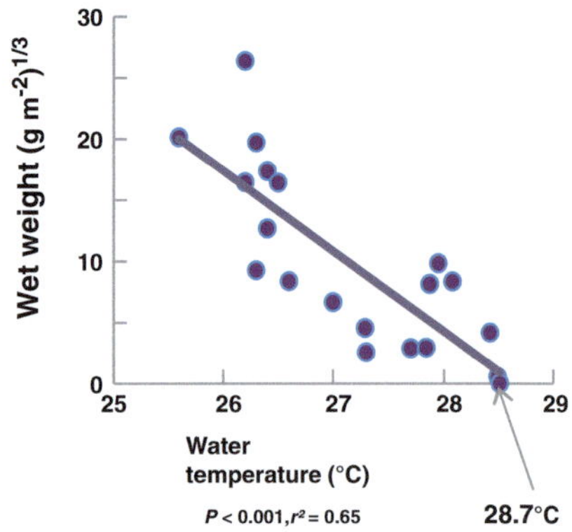

Fig. 4.33 Correlation between water temperature and the wet weight of the bivalve *Mytilus galloprovincialis* at a single sampling point and time in Dokai Bay. Samples were collected as described in Fig. 4.32, except no samples were used at the deeper sampling points affected by hypoxia

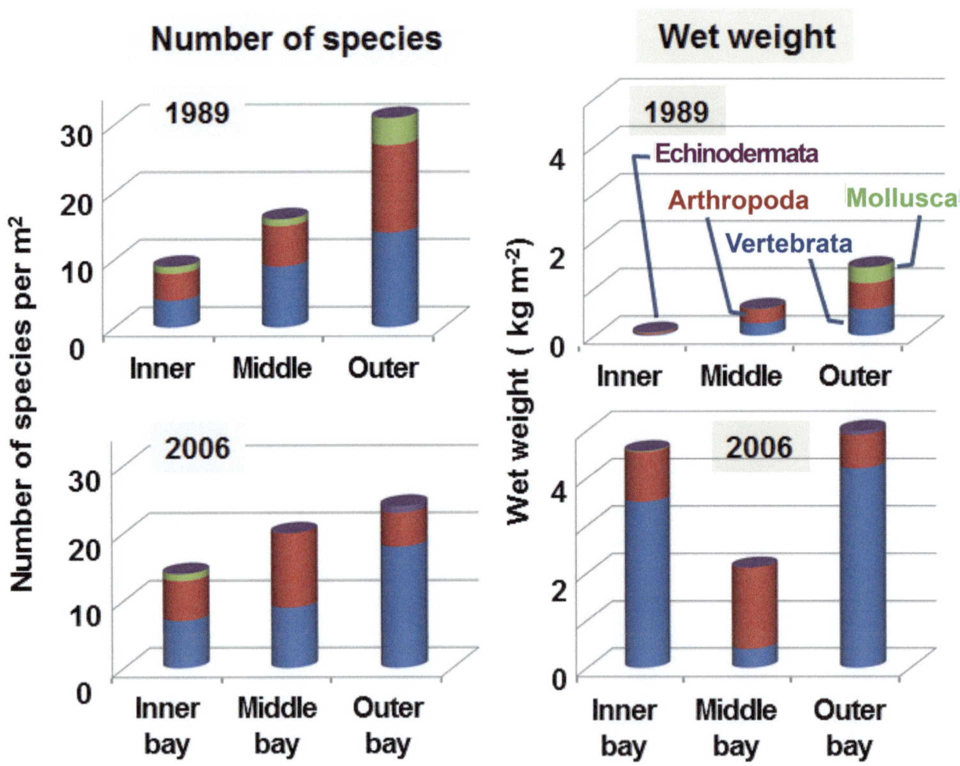

Fig. 4.34 Differences in the number of species and wet weight of fishes, shrimp, and crabs in the inner, middle, and outer parts of Dokai Bay during periods of high (August 1989) and low (August 2006) N and P concentrations

macrobenthos are collected from sediments of target sea areas and then cultured in land-based facilities; they are then scattered on the bottom sediments of the bay to grow and propagate (Tsutsumi et al. 1998). The fifth component is the management of nitrogen and phosphorus loading to prevent the formation of hypoxic conditions and red tides (Yanagi 1997, 1998).

4.4.2 Facility for Water Purification Using Filter-Feeding Bivalves

Bivalves, including short-necked clams, oysters, and mussels, are candidates for water purification organisms that can remove suspended particulate matters from seawater. Moreover, they are valuable as fishery resources. When

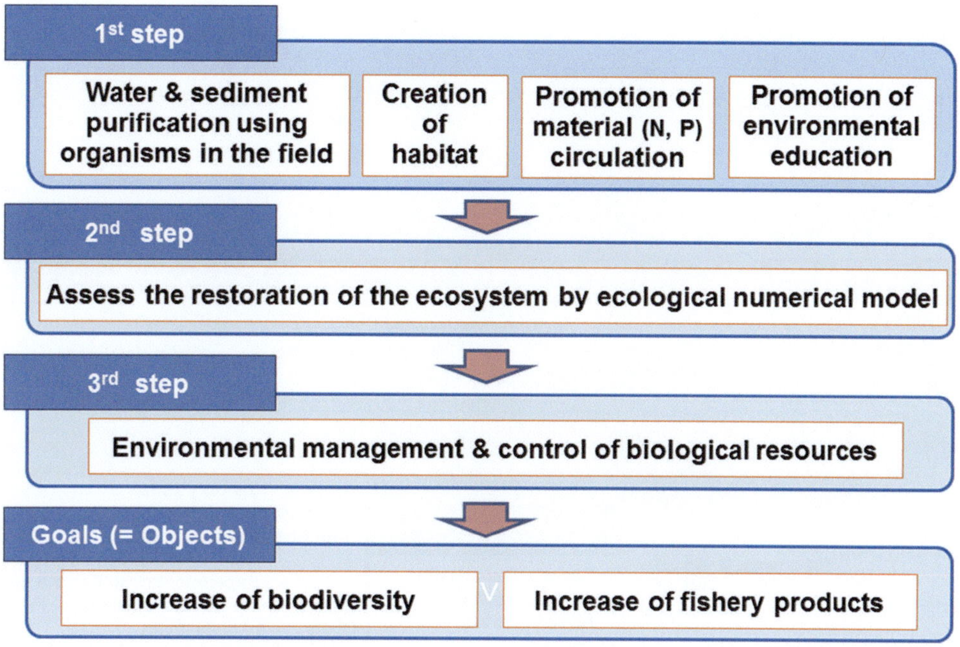

Fig. 4.35 Flow chart of the ecological environmental restoration method

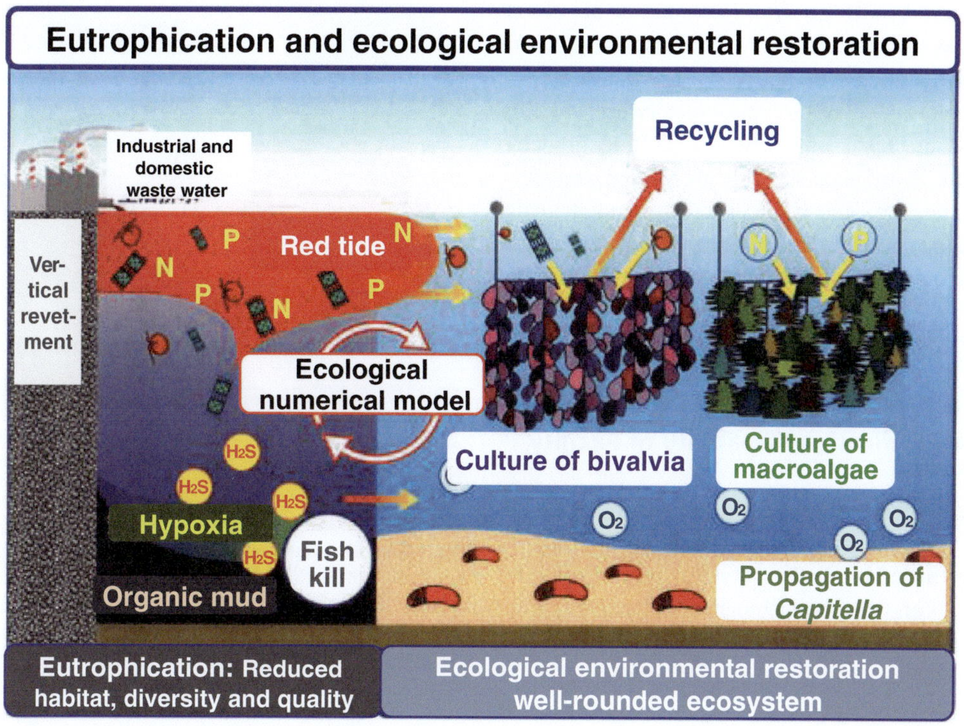

Fig. 4.36 Diagram showing the mechanism of eutrophication and the five components of ecological environmental restoration

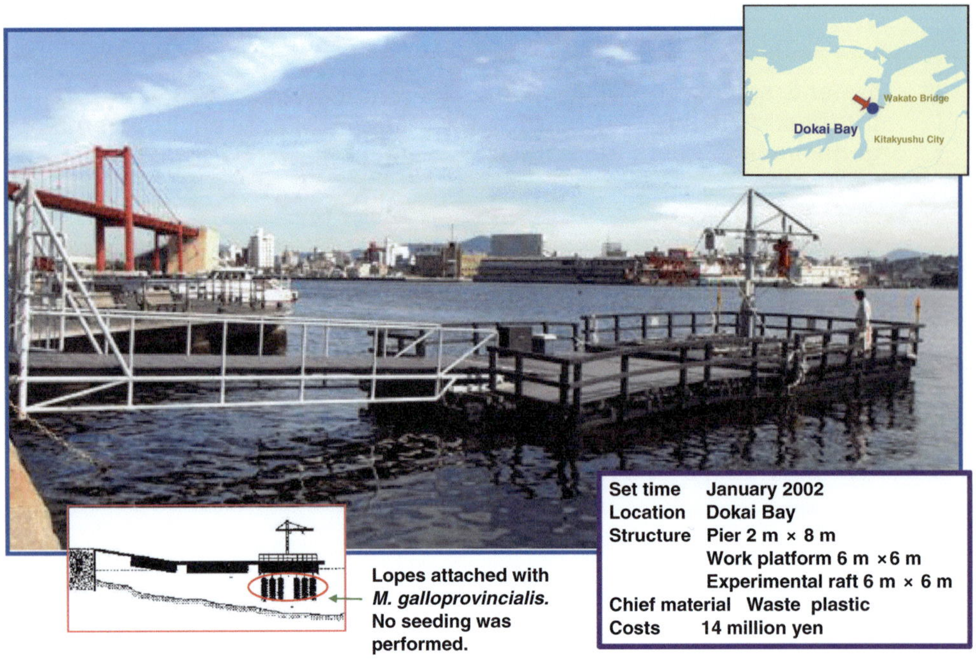

Fig. 4.37 A experimental restoration facility using the filter-feeding bivalve *Mytilus galloprovincialis* as purification organisms

we started the purification study in 2002, among these bivalves, only the mussel *Mytilus galloprovincialis* was commonly observed in Dokai Bay (Kohama et al. 2001). We therefore used this mussel as the purification organism even though it is an alien species.

An experimental environmental restoration facility was constructed in Dokai Bay (Fig. 4.37). Biodegradable plastic ropes with many filament loops were used as attachment sites for mussels. The ropes were suspended in the seawater of the bay in early spring, and about half a year later, they were collected, with plenty of attached mussels, for recycling (Fig. 4.38). The juvenile mussels attached to the ropes naturally by their byssus, so no seeding was necessary.

4.4.3 Water Purification Capability of Facility

Chl. *a* was significantly lower in the seawater inside the environmental restoration facility than outside of it (Fig. 4.39). We carried out both a physical investigation by surveying currents and tides and a biochemical investigation, in which we analyzed the concentrations of nitrogen nutrients and those in suspended and sinking particles over a period of 12 h. We measured the nitrogen uptake by this purification system to be as high as 940 mg m^{-2} in 1 h, a value 3–230 times the measured uptakes in organisms on/in tidal flat environments (Table 4.8). The high purification capability of this method was due to its utilization of three-dimensional space, not just flat surfaces such as those characteristic of water purification systems on tidal flats and quay walls.

Fig. 4.38 (**a**) A water purification organism (*M. galloprovincialis*). (**b**) Rope made of biodegradable plastic with many filament loops. (**c**) Top view of a rope with attached bivalves, collected after hanging for 4 months beneath the restoration facility. (**d**) Ropes with attached bivalves collected after half a year

4.4.4 Creation of Marine Biotope Using Facility

Different fish, including Black rockfish *Sebastes* spp., Sea bass *Lateolabrax japonicus*, Surfperch *Ditrema temmincki*, and Black sea bream *Acanthopagrus schlegelii*, were

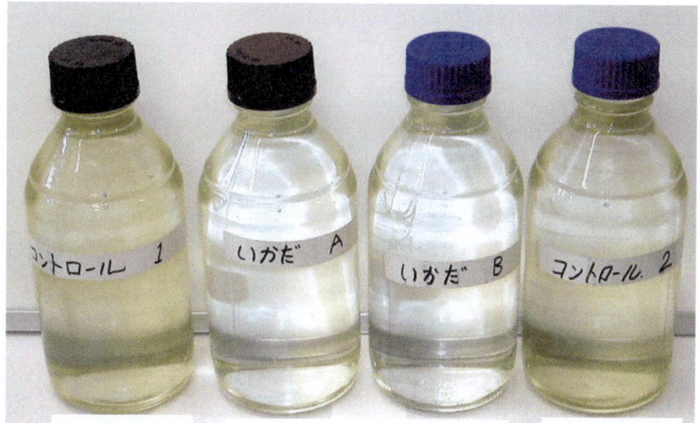

		Outside* Stn 1	Inside** Stn A	Inside** Stn B	Outside* Stn 2
Water temperature	°C	27.0	27.0	27.0	26.8
Salinity		29.2	29.3	29.2	29.4
Suspended solids	mg L^{-1}	30	4	13	32
Chl. *a*	mg L^{-1}	98	45	47	140

*, outside the facility; **, inside the facility

Fig. 4.39 Water purification capability of the experimental environmental restoration facility

Table 4.8 Comparison of nitrogen uptake by mussels in the experimental restoration facility with other water purification examples

Purification organism	Location	Uptake (mg m^{-2} h^{-1})
Mussel	Dokai Bay	940
M. galloprovincialis		
Short-neck clam[a]	Banzu tidal flat in Tokyo Bay	19–300
	Matsukawaura	12.6
Tidal flat system[a]	Mesocosm	3.0
	Banzu tidal flat in Tokyo Bay	6.9

[a]From Working Group of Nature Restoration in Coastal Waters 2003

observed to gather in, under, and around the experimental environmental restoration facility (Fig. 4.40). The number of fish in the vicinity of the facility was four times the number found near the quay wall and 20 times the number in a neighboring seawater area without any structures (Fig. 4.41). In addition, the number of fish observed in the vicinity of the facility increased around sunrise and sunset (Fig. 4.42). Thus, fish activity patterns were consistent with the reports of local fishermen that the fish become active around sunrise and sunset. These observational results showed that the fish gathered at the facility used it as a feeding site and/or for shelter, and their position in the food chain was confirmed by investigations of body length and stomach contents (Fig. 4.43). Thus, the restoration facility functions as a floating fish bank (Yamada et al. 2004).

Fig. 4.40 Fishes schooling in and under the experimental environmental restoration facility in Dokai Bay

4.4.5 Composting Using Filter-Feeding Bivalves

The mussels used as purification organisms do not have a place in Japanese dietary culture because they are a nonnative species. Therefore, we unfortunately had to compost them. To prevent the mussels from falling off the ropes because of hypoxia or from their own weight, the 1.5-m-long ropes were collected in August when their weight reached about 10 kg (Fig. 4.38). The collected mussels were then crushed and mixed with wood chips from pruned branches. The temperature of the mixture increased to above 70 °C within a few days and compost was successfully obtained within 3 months.

The usefulness of this mussel-wood chip compost was examined in comparison with compost made from wood chips alone and an ammonium sulfate-wood chip compost, in cultivation tests with Japanese mustard spinach *Brassica rapa* var. *perviridis*. The mussel-wood chip compost significantly stimulated the growth of the plants (Fig. 4.44). The nitrogen and phosphorus contents of the mussel-wood chip compost were higher than those of the compost made from wood chips only, and the mussel-wood chip compost also contained more humic acid, sodium, and calcium than the ammonium sulfate-wood chip compost. The concentrations of four hazardous components were under the limits set by the Japanese Fertilizer Control Act (2000) (Yamada et al. 2004).

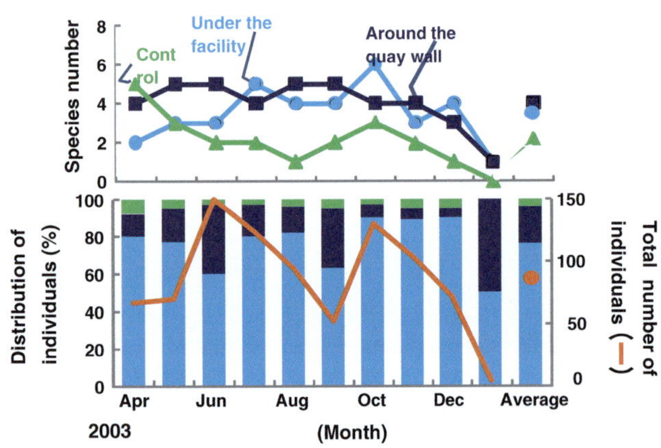

Fig. 4.41 Monthly changes in the total number of fish and the species composition of fishes observed at three locations using the fixed position method: in and under the experimental environmental restoration facility, in front of the neighboring quay wall, and in a seawater area without any structures (Control)

4.4.6 Environmental Education

The Seaport and Airport Bureau of Kitakyushu City has an educational program about the environment that targets elementary school students. Students in the program participate in "My rope" and "My compost" activities at the restoration facility (Fig. 4.45). The program, which has pre-learning, boarding practical works, and water quality measurement modules, is designed as a comprehensive introduction to eutrophication. More than 600 students participate each year in the program.

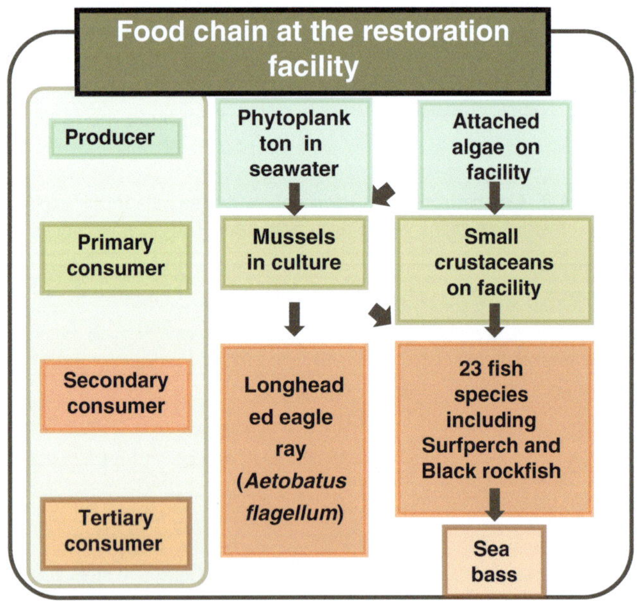

Fig. 4.43 Food chain at the experimental environmental restoration facility in Dokai Bay

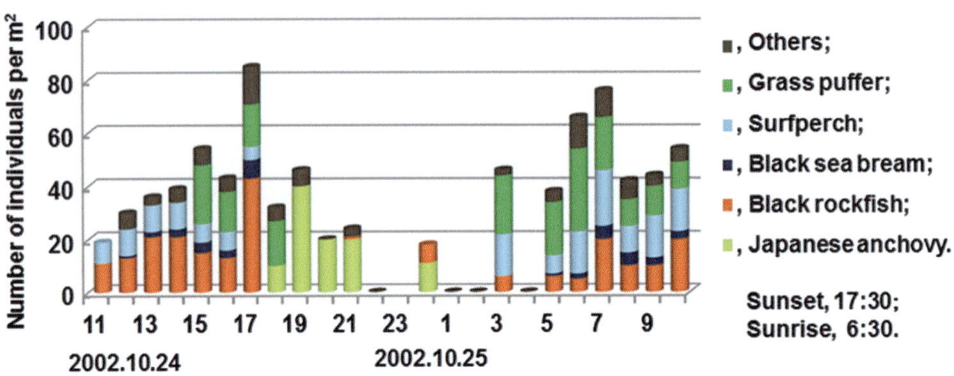

Fig. 4.42 Temporal changes of fish compositions in and under the experimental environmental restoration facility over a 24-h period

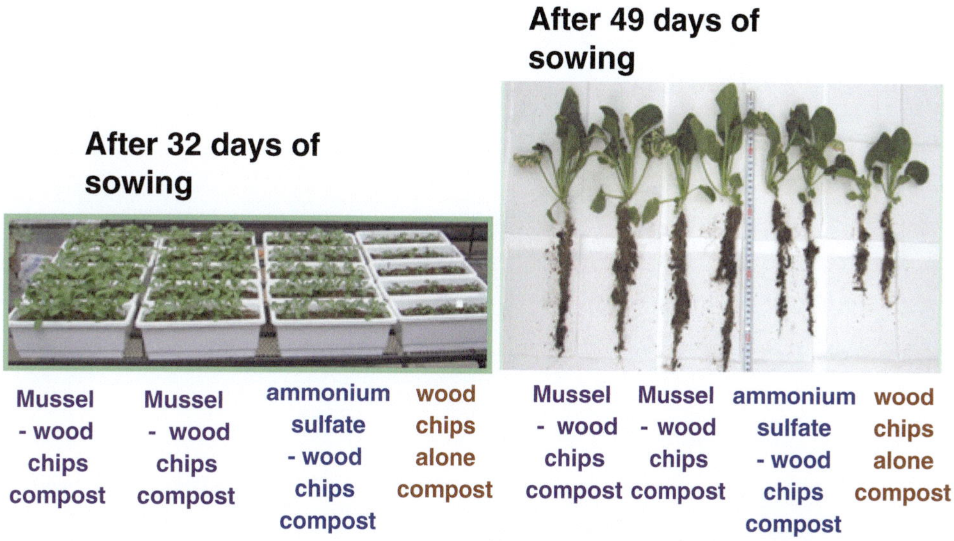

Fig. 4.44 Growth of Japanese mustard spinach *Brassica rapa* var. *perviridis* with three types of compost. Crushed mussels (*M. galloprovincialis*) and wood chips were used for the mussel-wood chips compost

Fig. 4.45 Student participation in the environmental learning program about eutrophication and the use of *M. galloprovincialis* as a purification organism. (**a**) Elementary school students hanging the ropes for mussels at the experimental environmental restoration facility. (**b**) Observing *M. galloprovincialis* in the laboratory

4.4.7 Restoration Using Macroalgae

The Seaport and Airport Bureau of Kitakyushu City created seaweed beds by constructing mounds with natural stones or iron and steel slag in 2008 (Fig. 4.46). In the first year, the mounds became covered by green algae in early spring, when macroalgae generally grow quickly, and the following year they became covered mainly by brown algae, including Japanese kelp *Undaria pinnatifida*, Jap weed *Sargassum muticum*, and Asian seaweed *Sargassum horneri*. More than 7 kg m^{-2} of brown algae was harvested from the mounds of iron and steel slag (Figs. 4.46 and 4.47).

Nineteen species of fish were observed to gather around the seaweed beds, among them are black rockfish *Sebastes* spp. as shown in Fig. 4.48. Large numbers of sea cucumbers *Apostichopus japonicus* were also observed around the beds, although we had not previously found sea cucumbers in Dokai Bay during our many biological investigations (Fig. 4.48). Sea cucumbers are a valuable fishery product, and the local loads at that time have exported dried sea cucumbers from Dokai Bay to China since the Edo period, more than 200 years ago.

4.5 Conclusive Words

Long-lasting hypoxia events were observed over large parts of Dokai Bay in the 1990s (Fig. 4.18), but the severity of the events and their duration decreased with the passage of time (Figs. 4.19 and 4.31). Dokai Bay was once ranked as one of the most damaged sea areas due to eutrophication after industrial pollution in Japanese coastal waters, with maximum total nitrogen and total phosphorus values of 18 mg L^{-1} and 0.43 mg L^{-1}, respectively (Fig. 4.14). Today, however, the water quality has improved sufficiently to meet the governmental environmental standards of 1 mg L^{-1} and 0.09 mg L^{-1}, respectively (Fig. 4.14). As a result of the discharge of industrial effluents, the concentration of NH_4-N was as high as 1,590 μM in 1995, high enough to kill phytoplankton, but today it has decreased to 13 μM. Although the cause of this successful decrease of nutrients has been inferred by the reduction of nitrogen and phosphorus loads, the exact mechanism has yet to be identified scientifically.

The improvement of various environmental factors has led to favorable changes in the eutrophic environment and

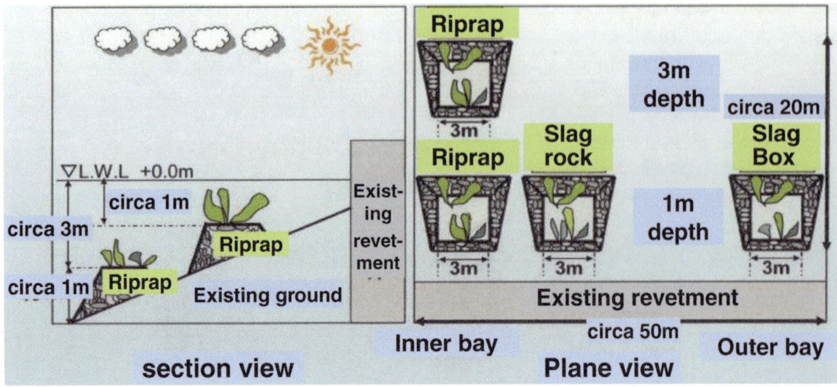

Fig. 4.46 Artificial mounds for seaweed beds in Dokai Bay

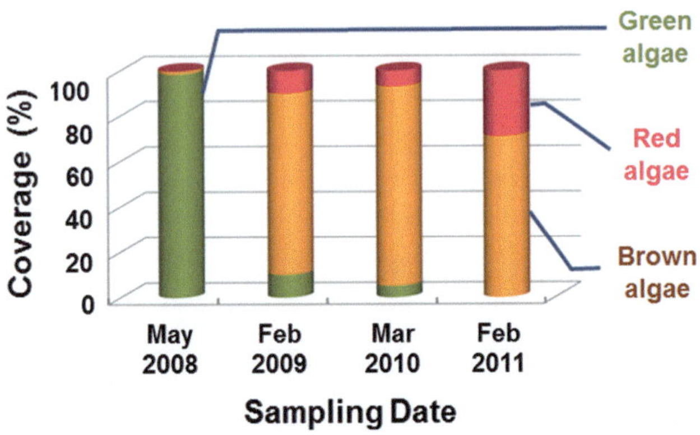

Fig. 4.47 Yearly changes in the macroalgae coverage (area covered by macroalgae relative to the total surface area of the mounds) on the artificial mounds

has altered the species composition of various taxa. Biological responses to the mitigation of eutrophic conditions in Dokai Bay are summarized in Table 4.9. The unusually high concentrations of NH_4-N during the 1990s inhibited the growth and proliferation of not only phytoplankton but also all other organisms (Shimo et al. 2004) in the bay and sometimes killed them (Fig. 4.49). As the concentrations of nutrients such as NH_4-N decreased, the species diversity of red tide organisms was observed to increase. The occurrence frequency of red tides, however, has not changed, so the reduction of nutrient levels did not cause the disappearance of red tides. The species diversity of macroalgae and the macroalgal biomass were observed to increase throughout the bay and vertically in the water

Fig. 4.48 Photographs of artificial mounds for seaweed beds in Dokai Bay. (**a–c**) Three species of brown algae *Undaria pinnatifida*, *Sargassum muticum*, and *Sargassum horneri* growing on the mounds. (**d**) Gathering with fish of several kinds. (**e**) Sea cucumber *Apostichopus japonicus* returning to the bay

Table 4.9 Biological responses to the improvement of eutrophication in Dokai Bay

Biota	Biodiversity	Biomass	Effects[a] and current status
Phytoplankton (red tide organisms)	Increased	No change	The decrease of nutrients may have increased the number of species forming red tides. Red tides, however, still form because nutrient concentrations have not been reduced enough to cause their disappearance
Macroalgae	Increased	Increased	Increased light intensity due to decreased eutrophication may have allowed macroalgae to grow in more parts of the bay and deeper in the water column
Sessile animals	Increased	Decreased	Raising the DO concentration in the bottom area of the inner bay allowed many kinds of sessile animals to populate this area. A rise in water temperature, however, may be inhibiting the growth of the most dominant species *M. galloprovincialis*
Fish, shrimp, and crabs	Increased	Increased	Raising the DO concentration in the bottom area of the inner bay has allowed many fish, shrimp, and crabs to recolonize the area

[a]High concentrations of NH_4-N in the 1990s commonly either inhibited the growth of various kinds of organisms or led to their death

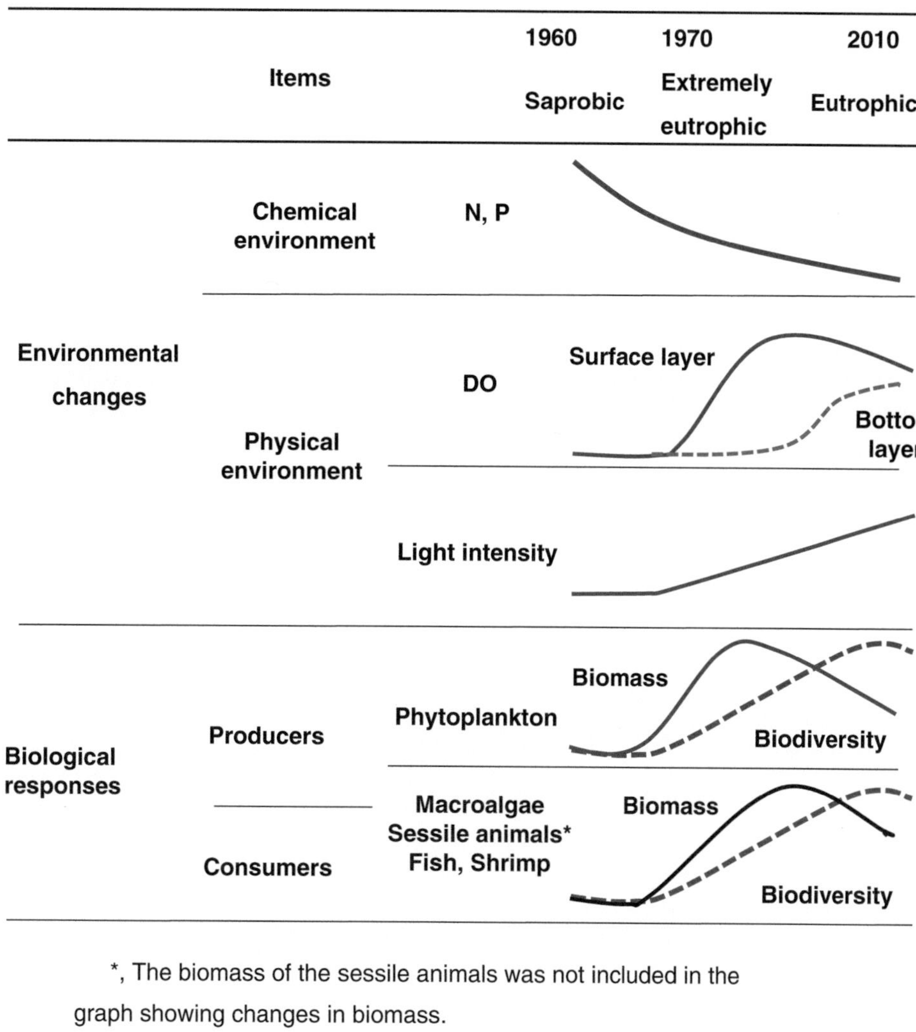

Fig. 4.49 Ecological improvement of the water environment of Dokai Bay during July and September from 1960 to 2010

column because of the increased penetration of light into the seawater as transparency improved.

The number of species of fish, shrimp, crabs, and sessile animals rose when dissolved oxygen levels at the bottom of the inner bay increased with the decreasing extreme eutrophication. In the case of fish, the biomass increased dramatically, whereas the biomass of sessile animals decreased markedly. The abrupt decrease of biomass, which was due to the disappearance of mussels, may have been due to the observed rise in summer water temperature. This notable decrease indicates that ecological restoration projects must take into account not only water pollution and eutrophication but also the impact of global warming.

In the 1960s, saprobic organisms lived on the excessive organic matter, the dissolved oxygen concentration was zero or near zero in all layers, and light penetration into the seawater was very weak, so the bay environment supported little besides anaerobic bacteria. At that time Dokai Bay was called a "dead sea," because it was severely affected by organic pollution and hazardous substances derived from industrial effluents.

During the 1970s to 1990s, the bay became extremely eutrophic because of the high nutrient loading in treated industrial effluents discharged into the bay. The oxygen content of the surface waters eventually improved as the amount of organic matter decreased. Surface waters often suffered heavy red tides composed of only a few phytoplankton species that were adapted to the high nutrient levels. Light penetration improved very little because of the severe and long-lasting red tides and high concentrations of suspended solids in industrial effluents. Therefore, the growth of perennial macroalgal and benthic microalgal producer species was suppressed. The occurrence of fish and sessile animals was also inhibited by the hypoxic conditions near the bottom of the bay.

In the 2000s, nutrient inflows decreased further, and light penetration increased because of the reduction of suspended matter, except phytoplankton, by the treatment of effluents. These changes led to an abundant macroalgal biomass. As the eutrophic level improved, the fish biomass also increased because of the abundant phytoplankton on which the fish could feed, and the species diversity of fish and sessile animals increased when hypoxic conditions no longer occurred. If nutrient levels continue to decrease and do not become a limiting factor on phytoplankton growth, the bay environment will return to its former favorable state and the biota will become more diverse. The results presented in this chapter show that the eutrophic level, especially severe eutrophication, has an adverse impact on the richness of the bay ecosystem and that species diversity and biomass both improve when severely eutrophic conditions are improved to a lower eutrophic level.

In addition, the construction of floating fish beds or seaweed beds in port and harbor areas, where tidal mud flats and shallow areas no longer exist, allows species previously found in the area to return in abundance. Therefore, environmental restoration projects enrich the ecosystems in the bay and contribute to the improvement of eutrophication.

If nutrient concentrations continue to decrease because of strict countermeasures, the biomass of phytoplankton and macroalgae, species at the base of the food chain, will decrease, which will lead in turn to a reduction in the biomass of valuable fishery products such as nori (edible seaweed), fish, and shrimp. An oligotrophic environment cannot supply the demand for fishery products because of the decrease in primary production. Biodiversity, biomass, the ecosystem, and ecosystem function change in response to the eutrophic level. It is a great challenge to determine what eutrophic level to choose in the future. Environmental management and biological resource controls will be required to bring the sea environment to the eutrophic level favored by citizens, along with anticipated advances in science and technology. Cooperation among government agencies, stakeholders, and specialists will be needed for adaptive management of Dokai Bay.

Box 4.2: Night View of Enormous Factories on Dokai Bay

Small cruising tours of the coastal landscapes at Dokai Bay are conducted for the purpose of environmental education. In addition, night view tours of the enormous factories on Dokai Bay were initiated in 2012. The motivation for these tours is the dramatic restoration of the bay's ecosystems, which had been damaged by serious industrial pollution and subsequent eutrophication due to excess nutrient loading (Yamada et al. 1990, 2000; Yanagi and Yamada 1999) (Fig. Box 4.2).

The two nighttime photographs of the factories are in stark contrast to their arid daytime appearance. The illumination of these factories, which is done for safety reasons, creates a strange aura of dynamic energy and grand beauty that attracts many tour participants.

Despite the economic importance of the factories and industrial activities around Dokai Bay, the heavy pollution that they caused in the past made them an unwelcome presence to citizens in the area. Now, however, these facilities have become a familiar, even attractive sight to most citizens. These factories illuminating the night have thus become a modern symbol of regional and urban development.

(continued)

Box 4.2 (continued)

Fig. Box 4.2 Night view of enormous factories on Dokai Bay

Acknowledgments I am grateful to Dr. Naoko Ueda, Dr. Ken-ichiro Hamada, and Mr. Tatsumi Murata of the University of Kitakyushu for providing recent useful biological and chemical data. I also thank Kitakyushu City workers and Mr. Hiroyuki Tajima and Mrs. Yuko Yano of the Kitakyushu Seaport and Airport Bureau for providing bioremediation data, Dr. Toshiro Yamashita of the Kitakyushu Environment Bureau for providing the data on nutrient loads, and Mrs. Akiko Terashi and Mrs. Yoko Kajiwara of the Institute of Environmental Sciences for skillful handling of the investigation of sessile animals. I especially thank the executive editor of "Estuaries in the World", Dr. Jean-Paul Ducrotoy and the joint authors of this book, Dr. Tetsuo Yanagi and Dr. Keita Furukawa, for constructive comments on the manuscript.

References

Central Environment Council, Ministry of the Environment (2010) Modalities of 7th water quality control by total quantity (Policy report), 66 p (in Japanese)

Food and Agriculture Organization of the United Nations (2005) Review of the state of world marine fishery resources, FAO fisheries technical paper no.5. FAO, Rome, v+235 p

Fukuoka Prefectural Fisheries Experiment Station (1933) Records on Dokai Bay, 136 p (in Japanese)

Hamada K, Ueda N, Yamada M, Tada K, Montani S (2010) Seasonal changes in biochemical characteristics and vertical transport of particulate organic matter in hypertrophic Dokai Bay, Japan. Jpn Bull Coast Oceanogr 48:75–85 (in Japanese with English abstract and captions)

Hamada K, Ueda N, Yamada M, Tada K, Montani S (2012) Decrease in anthropogenic nutrients and its effect on the C/N/P molar ratio of suspended particulate matter in hypertrophic Dokai Bay (Japan) in summer. J Oceanogr 68:173–182

Hashimoto S, Tada K (1997) Optical properties of seawater in Hiroshima Bay, Japan. Oceanogr Jpn 6:151–155 (in Japanese with English abstract and captions)

Heike KL, Flitner M, Urban ER Jr (2008) Coastal basins on the edge, Policy brief no. 7. UNESCO-SCOPE-UNEP, Paris, 6 p

Higashi T, Yamada M, Montani S, Hirotani J, Yanagi T (1998) Development and characteristics of oxygen-deficient water mass in hypertrophic Dokai Bay, Japan. Nippon Suisan Gakkaishi 64:204–210 (in Japanese with English abstract and captions)

Hulburt EM, Guillard RRL (1968) The relationship of the distribution of the diatom *Skeletonema tropicum* to temperature. Ecology 49:337–339

Inoue K, Yanagi T, Montani S, Yamada M (1998) Variation of residual flow in Dokai Bay. Mem Fac Eng Ehime Univ 17:283–290 (in Japanese with English abstract and captions)

Interagency Working Group on Harmful Algal Blooms, Hypoxia, and Human Health (2010) Scientific Assessment of Hypoxia in U.S. Coastal Waters, 105 p. http://www.whitehouse.gov/administration/eop/ostp/nstc/oceans

Japan meteorological Agency (2012) Meteorological statistical information. http://www.data.jma.go.jp/obd/stats/etrn/view/annually_a.php?prec_no=82&b

Kajiwara Y, Yamada M (1997) Study on occurrence characteristics of sessile animals and classification of eutrophic level in Dokai Bay. J Jpn Soc Water Environ 20:185–192 (in Japanese with English abstract and captions)

Kohama T, Montani S, Kajiwara Y, Yamada M (2001) Population dynamics of sessile bivalves *Mytilus galloprovincialis* and *Xenostrobus securis* in a hyper eutrophic bay, Japan. Nippon Suisan Gakkaishi 67:664–671 (in Japanese with English abstract and captions)

Montani S, Kohama T, Tokunaga Y, Yamada M (1998) The development of environmental restoration technology utilizing ecological characteristics of filter feeding Bivalvia. Environ Sci 11:407–420 (in Japanese)

Reid WV, Mooney HA, Cropper A, Capistrano D, Carpenter SR, Chopra K, Dasgupta P, Dietz T, Duraiappah AK, Hassan R, Kasperson R, Leemans R, May RM, McMichael T(AJ), Pingali P, Samper C, Scholes R, Watson RT, Zakri AH, Shidong Z, Ash NJ, Bennett E, Kumar P, Lee MJ, Raudsepp-Hearne C, Simons H, Thonell J, Zurek MB (2005) Ecosystems and human well-being: synthesis, millennium ecosystem assessment. Ed. by World Resources Institute, Island Press, Washington DC, p 137

Sarno D, Kooistra WHCF, Medlin LK, Percopo I, Zingone A (2005) Diversity in the genus *Skeletonema* (Bacillariophyceae). II. An assessment of the taxonomy of *S. costatum*-like species with the description of four new species. J Phycol 41:151–176

Sarno D, Kooistra WHCF, Hargraves PE, Zingone A (2007) Diversity in the genus *Skeletonema* (Bacillariophyceae). III. Phylogenetic position and morphology of *Skeletonema costatum* and *Skeletonema grevillei*, with the description of *Skeletonema ardens* sp. Nov. J Phycol 43:156–170

Shimo S, Akimoto Y, Takahama H (2004) Review of the water quality effects on marine organisms. Rep Mar Ecol Res Inst 6:1–159 (in Japanese)

Suksomjit M, Tada K, Funaki Y, Montani S, Yamada M, Harrison P (2009) Ammonium ion accelerates the growth rate of *Skeletonema* spp. in the phytoplankton assemblage in a heavily eutrophic embayment, Dokai Bay, Japan. La Mer 47:89–101

Tada K, Ichimi K, Yokota HT, Yamada M, Montani S (2004) Why flagellates do not produce bloom in Dokai Bay, Japan, Japan. J Oceanogr 13:271–279 (in Japanese with English abstract and captions)

Tada K, Ichimi K, Hamada K, Ueda N, Yamada M, Montani S (2007) Estuarine circulation and red tide outbreaks in Dokai Bay, Japan. Bull Coast Oceanogr 44:147–155 (in Japanese with English abstract and captions)

Takao T, Okada T, Nakayama K, Furukawa K (2004) Seasonal variation of residence time of seawater in Tokyo Bay during 2002. Technical note of National Institute for Land and Infrastructure Management, 169, pp 1–78 (in Japanese)

Tsutsumi H, Hamada K, Ueda N, Yamada M, Fujiki T, Nakamura H, Wada I, Tokuda T, Montani S (1998) Restoration of the sediment environment using *Capitella* sp. 1 Polychaete in the inner part of Dokai Bay. Environ Sci 11:421–429 (in Japanese)

Ueda N, Tsutsumi H, Yamada M, Takeuchi R, Kido K (1994) Recovery of the marine bottom environment of a Japanese bay. Mar Pollut Bull 23:676–682

Ueda N, Tsutsumi H, Yamada M, Hanamoto K, Montani S (2000) Impacts of oxygen-deficient water on the macrobenthic fauna of Dokai Bay, and on adjacent intertidal flats, in Kitakyushu, Japan. Mar Pollut Bull 40:906–913

Unoki S, Kishino M (1977) Currents and water exchange in Tokyo Bay. Technical report of physical oceanography laboratory in Institute of Physical and Chemical Research 1, pp 1–86 (in Japanese)

Working Group of Nature Restoration in Coastal Waters (2003) Handbook of nature restoration in coastal waters – design, technology and practice, vol 2, Tidal Flat edn. Gyosei, Tokyo, p. 31 (in Japanese)

World Resources Institute (2012) Interactive map of Eutrophication & Hypoxia. http://www.wri.org/project/eutrophication/map

Yamada M (2000) Chapter 12. Dokai Bay. In: Okada M, Peterson SA (eds) Water pollution control, policy and management: The Japanese experience. Gosei, Tokyo, pp 61–173

Yamada M, Takeuchi R, Sueta S, Kido K, Yabumoto Y, Yoshida Y (1990) Recovery of aquatic animals in Dokai Bay, Northern Kyushu, Japan. Mar Pollut Bull 23:201–207

Yamada M, Higashi T, Hamada K, Ueda N, Eguchi N, Suzuki M (1998) Environmental improvement and management with an ecological environmental restoration method. Environ Sci 11:393–399 (in Japanese)

Yamada M, Tanaka K, Yoshikawa H (2004) Present condition of aquatic environment and eco-remediation projects in Dokai Bay, Kitakyushu City. J Environ Lab Assoc 29:95–101 (in Japanese)

Yamada M, Ueda N, Hanada Y (2005) Study on occurrence characteristics of seaweeds and eutrophic level of Dokai Bay, Kitakyushu City. J Environ Lab Assoc 30:252–258 (in Japanese)

Yamada M, Katsuki E, Otsubo M, Hamada K, Ueda N, Montani S (2009) Survival strategy of tropical and subtropical marine diatom *Skeletonema tropicum* in temperate coastal small bay (Dokai Bay), Japan. Oceanogr Jpn 18:157–167 (in Japanese with English abstract and captions)

Yamada M, Katsuki E, Otsubo M, Kawaguchi M, Ichimi K, Tada K, Harrison PJ (2010) Species diversity of the genus *Skeletonema* (Bacillariophyceae) in the industrial harbor Dokai Bay, Japan. J Oceanogr 66:755–771

Yamada M, Ueda N, Hamada K (2011) Changes in red tide occurrence and organisms responsible with declining eutrophic level in hypereutrophic Dokai Bay, Japan. Nippon Suisan Gakkaishi 77:647–655

Yanagi T (1997) Ecological modeling as tool for coastal zone management in Dokai Bay, Japan. J Mar Syst 13:123–136

Yanagi T (1998) The management of coastal waters and a numerical ecosystem model. Environ Sci 11:381–391 (in Japanese)

Yanagi T, Takahashi S (1988) Response to freshwater discharge in Osaka Bay. Sea Sky 64:63–70 (in Japanese with English abstract and captions)

Yanagi T, Yamada M (2000) Why red tides do not occur during winter in Dokai Bay, Japan. J Oceanogr 9:125–132 (in Japanese with English abstract and captions)

Yanagi T, Yamada M, Suzuki (1999) A challenge of water purification in Dokai Bay, Japan. Mar Pollut Bull 38:1063–1069

Yanagi T, Yamada M, Nakajina M (2001) Comparison of eutrophication mechanisms of Dokai and Hakata bays. J Oceanogr 10:275–283 (in Japanese with English abstract and captions)

Zingone A, Percopo I, Sims PA, Sarno D (2005) Diversity in the genus *Skeletonema* (Bacillariophyceae). I. A reexamination of the type material of *S. costatum* with the description of *S. grevillei* sp. nov. J Phycol 41:140–150

MIX
Papier aus verantwortungsvollen Quellen
Paper from responsible sources
FSC® C105338
FSC
www.fsc.org

If you have any concerns about our products,
you can contact us on
ProductSafety@springernature.com

In case Publisher is established outside the EU,
the EU authorized representative is:
Springer Nature Customer Service Center GmbH
Europaplatz 3, 69115 Heidelberg, Germany

Printed by Libri Plureos GmbH
in Hamburg, Germany